Solutions Manual

Introduction to Counting and Probability
2nd edition

David Patrick
Art of Problem Solving

Art of Problem Solving®

Books • Online Classes • Videos • Interactive Resources

www.artofproblemsolving.com

Published by: AoPS Incorporated
                 10865 Rancho Bernardo Rd Ste 100
                 San Diego, CA 92127-2102
                 books@artofproblemsolving.com

ISBN: 978-1-934124-11-6

Visit the Art of Problem Solving website at http://www.artofproblemsolving.com

 Scan this code with your mobile device to visit the Art of Problem Solving website, to view our other books, our free videos and interactive resources, our online community, and our online school.

Cover image designed by Vanessa Rusczyk using KaleidoTile software.

Satellite image of Corral de Piedra, Argentina from NASA Earth Observatory.

Printed in the United States of America.

Second Edition. Printed in 2019.

Foreword

This book contains the full solution to every Exercise, Review Problem, and Challenge Problem in the text *Introduction to Counting and Probability*.

In most problems, the final answer is contained in a box, $\boxed{\text{like this}}$. However, we strongly recommend against just looking up the final answer and moving on to the next problem. Instead, even if you got the right answer, read the solution in this book. It might show you a different way of solving the problem that you might not have thought of.

If you don't understand a solution, or you think you have a better way of solving the problem, or (gasp!) find an error in one of our solutions, we invite you to come to our message board at

www.artofproblemsolving.com

and discuss it. Our message board is free to use and includes thousands of the world's most eager mathematical problem-solvers.

Contents

1

Counting Is Arithmetic

Exercises for Section 1.2

1.2.1 We subtract 35 from each member of the list to get $1, 2, 3, \ldots, 57, 58$, so there are $\boxed{58}$ numbers. Note that this follows the $b - a + 1$ formula for how many numbers there are between a and b inclusive, as $93 - 36 + 1 = 58$.

1.2.2 Dividing each member of the list by 2, we get $2, 3, 4, \ldots, 64, 65$, and then subtracting 1, we get $1, 2, 3, \ldots, 63, 64$, so there are $\boxed{64}$ numbers.

1.2.3 Add 3 to each member of the list to get $-30, -25, -20, \ldots, 55, 60$, and divide by 5 to get $-6, -5, -4, \ldots,$ $11, 12$. Using the inclusive integer formula, we get $12 - (-6) + 1 = 19$, so there are $\boxed{19}$ numbers. We could also add 7 to each number in the list to get $1, 2, 3, \ldots, 18, 19$, and see that there are 19 numbers.

1.2.4 First, we reverse the list to become $39, 42, \ldots, 144, 147$. Then we divide each number by 3 to get $13, 14, \ldots, 48, 49$, so there are $49 - 13 + 1 = \boxed{37}$ numbers.

1.2.5 Multiplying each number by 3, we get $11, 13, 15, \ldots, 79, 81$. Then we can subtract 1 and divide by 2 to get $5, 6, 7, \ldots, 39, 40$. So there are $40 - 5 + 1 = \boxed{36}$ numbers.

1.2.6 $7 \times 21 = 147 < 150 < 154 = 7 \times 22$, so $\boxed{21}$ positive multiples of 7 are less than 150.

1.2.7 Since $7^2 < 50 < 8^2$ and $15^2 < 250 < 16^2$, the squares between 50 and 250 are $8^2, 9^2, 10^2, \ldots, 15^2$. So there are $15 - 8 + 1 = \boxed{8}$ such squares.

1.2.8 Since $1^2 < 5 < 3^2$ and $13^2 < 211 < 15^2$, we have the list $3^2, 5^2, 7^2, \ldots, 13^2$, which has the same number of elements as $3, 5, 7, \ldots, 13$, which has $\boxed{6}$ elements.

1.2.9 Note that $17^4 = 83{,}521 < 100{,}000 < 104{,}976 = 18^4$. Since $17.5^4 \approx 16 \times 17 \times 18 \times 19$, we check the value of $16 \times 17 \times 18 \times 19 = 93{,}024$. Also $17 \times 18 \times 19 \times 20 = 116{,}280$, so $16 \times 17 \times 18 \times 19$ is the largest product of four consecutive positive integers which is less than 100,000. So there are $\boxed{16}$ sets.

Exercises for Section 1.3

1.3.1 Let the number of red 4-door cars be x. Since there are 12 red cars and 15 4-door cars, the number of red 2-door cars is $12 - x$, while the number of white 4-door cars is $15 - x$. The sum of the number of red 4-doors, red 2-doors, white 4-doors, and white 2-doors is the total number of cars (20), because each car is contained in exactly one of these categories. Since the number of white 2-doors is 4, we have $x + (12 - x) + (15 - x) + 4 = 20$, which makes $x = \boxed{11}$.

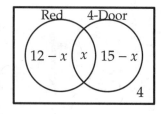

1.3.2 7 players are taking biology, so $12 - 7 = 5$ players are not taking biology, which means 5 players are taking chemistry alone. Since 2 are taking both, there are $5 + 2 = \boxed{7}$ players taking chemistry.

1.3.3 Let the number of blue-eyed students be x, so the number of blond students is $2x$. Since the number of blue-eyed blond students is 6, the number of blue-eyed non-blond students is $x - 6$, while the number of blond non-blue-eyed students is $2x - 6$. Since the number of non-blue-eyed non-blond students is 3, we can add up these four exclusive categories (blond blue-eyed, blond non-blue-eyed, etc.) to sum to 30 students in the class. So $(x-6)+(2x-6)+6+3 = 30$ and $x = \boxed{11}$, which is the number of blue-eyed students.

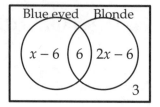

1.3.4 We draw a Venn Diagram with three circles, and fill it in starting with the center and proceeding outwards. There are 9 dogs that can do all three tricks. Since 18 dogs can sit and roll over (and possibly stay) and 9 dogs can sit, roll over, and stay, there are $18 - 9 = 9$ dogs that can sit, roll over, but not stay. Using the same reasoning, there are $12 - 9 = 3$ dogs that can stay, roll over, but not sit, and $17 - 9 = 8$ dogs that can sit, stay, but not roll over.

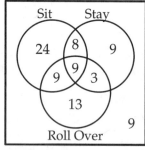

So now we know how many dogs can do multiple tricks, and exactly what tricks they can do. Since 50 dogs can sit, 9 dogs can sit and roll over only, 8 dogs can sit and stay only, and 9 dogs can do all three tricks, the remaining dogs that can't do multiple tricks can only sit, and there are $50 - 9 - 8 - 9 = 24$ of these. Using the same reasoning, we find that $29 - 3 - 8 - 9 = 9$ dogs can only stay and $34 - 9 - 3 - 9 = 13$ dogs can only roll over.

Since 9 dogs can do no tricks, we can add each category in the Venn Diagram to find that there are a total of $9 + 9 + 3 + 8 + 24 + 13 + 9 + 9 = \boxed{84}$ dogs and $8 + 9 + 3 = \boxed{20}$ dogs that can do exactly two tricks.

1.3.5 Since x people are in French and z are in both, $x - z$ are only in French. Similarly, $y - z$ are only in Spanish. Everyone in the school is either in French only, Spanish only, or both, so the total number of people in the school is $(x - z) + (y - z) + z = \boxed{x + y - z}$.

Exercises for Section 1.4

1.4.1 There are 6 choices of shirt and 5 choices of tie, so the total number of outfits is $6 \times 5 = \boxed{30}$.

1.4.2 There are 8 options for the shirt, and only 7 options for the tie because one of the ties has the same color as the shirt, so the number of outfits is $8 \times 7 = \boxed{56}$.

1.4.3 There are 26 choices of letters for each of the first two spots, and 10 choices of digits for each of the next 3, for a total of $26^2 \times 10^3 = \boxed{676,000}$ different plates.

1.4.4 There are 26 choices of letters for each of the first three spots, and 5 choices of digits for each of the last four spots (5 even or odd digits). This gives a total of $26^3 \times 5^4 = \boxed{10,985,000}$.

1.4.5 There are 5 options for the bottom book, 4 remaining options for the next book, 3 remaining options for the next book, 2 remaining options for the fourth book, and finally only 1 option for the top book. This gives a total of $5 \times 4 \times 3 \times 2 \times 1 = \boxed{120}$ ways to stack 5 books.

1.4.6 We first place the math books. We have two choices for the bottom book, and then the only one remaining choice for the top book which is the other math book. Then we place the four other books in the middle. There are 4 choices for the first book, 3 choices for the second, 2 choices for the third, and only 1 choice for the fourth. So the total number of ways the books can be placed is $2 \times 1 \times 4 \times 3 \times 2 \times 1 = \boxed{48}$.

1.4.7 There are 8 possible sprinters to award the gold to, then 7 remaining sprinters for the silver, and then 6 left for the bronze, for a total of $8 \times 7 \times 6 = \boxed{336}$ ways to award the medals.

1.4.8 There are 15 choices for president, 14 choices for vice-president, 13 choices for secretary, and 12 choices for treasurer, for a total of $15 \times 14 \times 13 \times 12 = \boxed{32,760}$ different choices.

1.4.9

(a) $9!/8! = \dfrac{9 \times 8 \times 7 \times 6 \times \cdots \times 1}{8 \times 7 \times 6 \times \cdots \times 1} = \boxed{9}$.

(b) $42!/40! = \dfrac{42 \times 41 \times 40 \times 39 \times \cdots \times 1}{40 \times 39 \times \cdots \times 1} = 42 \times 41 = \boxed{1,722}$.

(c) $8! - 7! = 8 \times 7! - 7! = 7!(8 - 1) = 7! \times 7 = 5040 \times 7 = \boxed{35,280}$.

Exercises for Section 1.5

1.5.1

(a) $P(8,3) = \dfrac{8!}{(8-3)!} = \dfrac{8!}{5!} = 8 \times 7 \times 6 = \boxed{336}$.

(b) $P(20,4) = \dfrac{20!}{(20-4)!} = \dfrac{20!}{16!} = 20 \times 19 \times 18 \times 17 = \boxed{116,280}$.

(c) $P(30,1) = \dfrac{30!}{(30-1)!} = \dfrac{30!}{29!} = \boxed{30}$.

(d) $P(6,5) = \dfrac{6!}{(6-5)!} = \dfrac{6!}{1!} = 6 \times 5 \times 4 \times 3 \times 2 \times 1 = \boxed{720}$.

(e) $P(50,3) = \dfrac{50!}{(50-3)!} = \dfrac{50!}{47!} = 50 \times 49 \times 48 = \boxed{117,600}$.

1.5.2 $P(n,n) = \dfrac{n!}{(n-n)!} = \dfrac{n!}{0!} = \boxed{n!}$. This is the number of permutations of size n from a group of n

objects, or just the number of ways to order n objects.

1.5.3

(a) There are 12 choices for the first ball, 11 remaining choices for the second ball, and 10 remaining choices for the third ball, for a total of $12 \times 11 \times 10 = \boxed{1320}$ possible drawings. Note that this is also just $P(12, 3)$.

(b) There are 12 options for each ball to be drawn, so there are a total of $12^3 = \boxed{1728}$ possible drawings.

(c) There are 12 options for the first ball, 12 options for the second ball (since it is replaced), and 11 options for the third ball (since the second ball is not replaced), for a total of $12 \times 12 \times 11 = \boxed{1584}$ possible drawings.

1.5.4 Using algebra, we see that

$$P(n, k)P(n - k, j) = \frac{n!}{(n - k)!} \times \frac{(n - k)!}{(n - k - j)!} = \frac{n!}{(n - k - j)!} = \boxed{P(n, k + j)}.$$

We are counting the number of permutations of size k from a group of n objects and then multiplying by the number of permutations of size j from the remaining $n - k$ objects. If we combine these two steps, it is the same as the number of permutations of size $k + j$ from the original n objects.

Review Problems

1.17 We can add 0.5 to each member of the list, to make it easier to deal with:

$$3, 6, 9, 12, \ldots, 81, 84.$$

Now if we divide by 3, we get

$$1, 2, 3, 4, \ldots, 27, 28,$$

so there are $\boxed{28}$ numbers in the list.

1.18 Subtract 2 from the list to get $4, 8, 12, \ldots, 80, 84$, and then divide by 4 to get $1, 2, 3, \ldots, 20, 21$. So the list has $\boxed{21}$ numbers.

1.19 Note that $7 \times 14 = 98 < 100 < 105 = 7 \times 15$ and $7 \times 142 = 994 < 1000 < 1001 < 7 \times 143$. So the list of 3-digit numbers divisible by 7 is $105, 112, \ldots, 994$, and when we divide this list by 7, we get the list $15, 16, 17, \ldots, 141, 142$, which has $142 - 15 + 1 = \boxed{128}$ numbers.

1.20 Let the number of lefty jazz lovers be x. So $8 - x$ lefties dislike jazz and $15 - x$ jazz lovers are righties. Since the number of righty jazz dislikers is 2 and the total number of members of the club is 20, we can add these four exclusive categories to get $x + (8 - x) + (15 - x) + 2 = 20$, so $x = \boxed{5}$, which is the number of lefty jazz lovers.

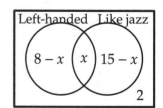

4

1.21 We can solve this using basic arithmetic: If 13 of the students with calculators are girls, and 22 students total have calculators, then $22-13 = 9$ of the students with calculators are boys. So if 9 boys have calculators, and there are 16 boys total, then $16-9 = \boxed{7}$ boys don't have calculators. Alternatively, we could solve this problem using a Venn diagram—the completed Venn diagram is shown at right. (Note that we don't have enough information to determine how many girls didn't bring their calculators.)

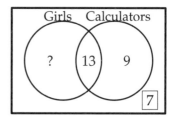

1.22 There are 5 options for shirts, 6 options for pants, and 8 options for hats, for a total of $5\times6\times8 = \boxed{240}$ outfits.

1.23 There are 26 options for the first letter, only 5 options for the second (vowels only), and 25 options for the third letter (all letters but the first letter). This gives $26 \times 5 \times 25 = \boxed{3{,}250}$ combinations.

1.24 We must count the number of permutations of 6 people. There are 6 choices for the first person in line, 5 choices for the second person in line, etc. So the answer is $\boxed{6! = 720}$.

1.25 There are 7 options for the first hat, 6 options for the second hat, etc. So the answer is $\boxed{7! = 5{,}040}$.

1.26 5 players are being chosen in order out of 12, so the answer is $P(12,5) = 12!/7! = 12\times11\times10\times9\times8 = \boxed{95{,}040}$.

1.27

(a) Obviously the only arrangement is T.

(b) There are two arrangements: IT and TI. One starts with I and one starts with T.

(c) There are 6 arrangements of ETI: $\boxed{EIT}$, $\boxed{ETI}$, IET, ITE, TEI, and TIE. Two of them start with E, and notice that the last two letters of those give the 2 arrangements of IT from part (b).

(d) There are $4! = 24$ arrangements of SETI: EIST, EITS, ESIT, ESTI, ETIS, ETSI, IEST, IETS, ISET, ISTE, ITES, ITSE, $\boxed{\text{SEIT, SETI, SIET, SITE, STEI, STIE}}$, TEIS, TESI, TIES, TISE, TSEI, TSIE. The 6 circled arrangements start with S. Note that the last 3 letters of each of the circled arrangements are exactly the 6 arrangements of ETI from part (c).

(e) We can think of this as 5 choices for the first letter, then $4! = 24$ arrangements for the other four letters, for a total of $5 \times 4! = 5! = \boxed{120}$ arrangements.

1.28

(a) The voting options are given by the following tree:

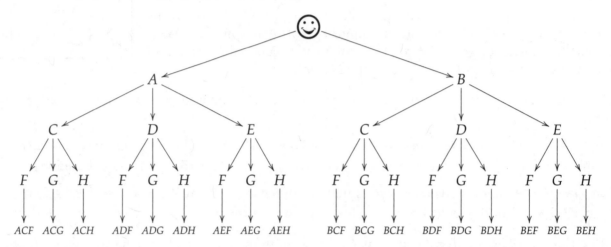

(b) There are 2 choices for the President, 3 choices for the Vice-President, and 3 choices for the Secretary, so there are a total of $2 \times 3 \times 3 = \boxed{18}$ ways to fill out a ballot. Notice that this corresponds to the 18 options at the bottom of our tree from part (a).

(c) Now there are 3 choices for the President (Abe, Barb, or no one), 4 choices for the Vice-President, and 4 choices for the Secretary, so there are a total of $3 \times 4 \times 4 = \boxed{48}$ ways to fill out a ballot.

Challenge Problems

1.29 Note that $7^3 < 500 < 8^3$, so any positive integer that can be written as the sum of two positive perfect cubes must be written as the sum of two cubes $a^3 + b^3$ where $1 \le a \le 7$ and $1 \le b \le 7$. We can make a chart of the sum of two such cubes, where those sums over 500 are crossed out:

	1^3	2^3	3^3	4^3	5^3	6^3	7^3
1^3	2	9	28	65	126	217	344
2^3		16	35	72	133	224	351
3^3			54	91	152	243	370
4^3				128	189	280	407
5^3					250	341	468
6^3						432	~~559~~
7^3							~~686~~

As we can see from the chart, there are $\boxed{26}$ such numbers.

1.30 Let x be number of people wearing all three items. Since 30 people are wearing bathing suits and sunglass, we know that $30 - x$ are wearing just bathing suits and sunglasses. Similarly, $25 - x$ are wearing just bathing suits and hats, while $40 - x$ are wearing just sunglasses and a hat.

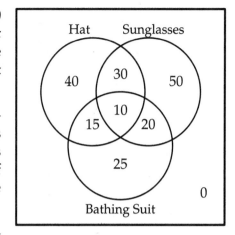

To find the number of people wearing just sunglasses, we subtract the people who are wearing sunglasses with other items from the total number of people wearing sunglasses, which is $110 - (30 - x) - (40 - x) - x = 40 + x$. Similarly, the number of people wearing just hats is $30 + x$, while the number of people wearing just bathing suits is $15 + x$.

Since the total number of people on the beach is 190, and everyone is wearing one of the items, we have:

$$190 = (15 + x) + (40 + x) + (30 + x) + (25 - x) + (30 - x) + (40 - x) + x = 180 + x.$$

We can then solve for x, so the number of people on the beach wearing all three items is $x = \boxed{10}$. You should check that this gives all of the numbers in the Venn diagram shown above.

1.31 Let x be the number of students taking physics, so the number in chemistry is $2x$. There are 15 students taking all three, and 30 students in both physics and calculus, meaning there are $30 - 15 = 15$ students in just physics and calculus. Similarly there are 60 students in just chemistry and calculus, and 60 in physics and chemistry. Since there are x students in physics and $15 + 15 + 60 = 90$ students taking physics along with other classes, $x - 90$ students are just taking physics. Similarly, there are $2x - 135$ students taking just chemistry and 90 students taking just calculus. Knowing that there are 15 students not taking any of them, the sum of these eight categories is 360, the total number of people at the school:

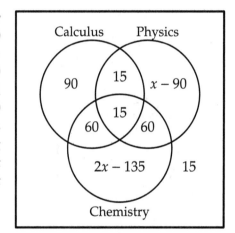

$$(x - 90) + (2x - 135) + 90 + 60 + 15 + 60 + 15 + 15 = 360.$$

We solve for x and find that the number of physics students is $x = \boxed{110}$.

1.32 Let the number of middle school boys be x. Since there are twice as many boys in high school than in middle school, the number of high school boys is $2x$. Since there are three times as many girls in middle school than boys in middle school, the number of middle school girls is $3x$. Since this represents half the girls in the club, then the number of high school girls is also $3x$. So the total number of the people in the club is $72 = x + 2x + 3x + 3x$, which means the number of middle school boys is $x = \boxed{8}$.

1.33 Note that 5! divides 10! and 5! divides 15!. Since 5! has no factor larger than 5!, and 5! is a factor of all three, the answer is $\boxed{5!}$.

1.34 The units digit of 1! is 1, the units digit of 2! is 2, the units digit of 3! is 6, the units digit of 4! = 24 is 4, and the units digit of 5! = 120 is 0. For all $n \geq 5$, $n!$ is a multiple of 5!, which is a multiple of 10, so all for all $n \geq 5$, the units digit of $n!$ is 0. This means that the units digit of the sum $1! + 2! + 3! + 4! + 5! + \cdots + 1000!$ is just the units digit of $1 + 2 + 6 + 4 + 0 + \cdots + 0 = 13$, so the answer is $\boxed{3}$.

1.35 To have 9 as a factor, $n!$ must have two factors of 3. The first such n for which this is true is 6, since $6! = \mathbf{6} \times 5 \times 4 \times \mathbf{3} \times 2 \times 1$. Since 9 is a factor of $6!$ and $6!$ is a factor of $n!$ for all $n \geq 6$, the numbers $6!, 7!, 8!, \ldots, 99!, 100!$ are all divisible by 9. There are $100 - 6 + 1 = \boxed{95}$ numbers in that list.

1.36 Multiplying the equation by $100n$, we get $50n > 100 > 3n$. Since $50n > 100$, we have $n > 2$, and since $100 > 3n$, we have $\frac{100}{3} > n$. The integers that satisfy both these inequalities are $3, 4, 5, \ldots, 32, 33$, and there are $33 - 3 + 1 = \boxed{31}$ numbers in this list.

1.37

(a) Each row has odd-numbered chairs $1, 3, 5, 7, 9, 11$ for a total of 6 odd-numbered chairs in each row. Since there are 11 rows, there are a total number of $11 \times 6 = \boxed{66}$ chairs with odd numbers.

(b) We separate this problem into two cases: n is odd or n is even.

   If n is odd, each row has odd-numbered chairs $1, 3, 5, \ldots, n-2, n$. Adding 1 to this list and dividing by two, we get $1, 2, 3, \ldots, \frac{n-1}{2}, \frac{n+1}{2}$. So there are $\frac{n+1}{2}$ odd-numbered chairs in each row and n rows, for a total of $\boxed{\frac{n(n+1)}{2}}$.

   If n is even, each row has odd-numbered chairs $1, 3, 5, \ldots, n-3, n-1$. Adding 1 to this list and dividing by two, we get $1, 2, 3, \ldots, \frac{n-2}{2}, \frac{n}{2}$. So there are $\frac{n}{2}$ odd-numbered chairs in each row and n rows, for a total of $\boxed{\frac{n^2}{2}}$.

1.38 Notice that there are 21 rows of dots and 11 columns of dots. Since there are 20 vertical toothpicks in each column and 11 columns, there are $20 \times 11 = 220$ vertical toothpicks. Similarly, since there are 10 horizontal toothpicks in each row and 21 rows, there are $21 \times 10 = 210$ horizontal toothpicks. This gives a total of $220 + 210 = \boxed{430}$ toothpicks.

CHAPTER 2

Basic Counting Techniques

Exercises for Section 2.2

2.2.1 There are three cases.

Case 1: The 3-letter word contains one A. The letter A can be in the 1st, 2nd, or 3rd position. Each of the other two positions can be one of B, C, or D. There are $3 \times 3 \times 3 = 27$ words for this case.

Case 2: The 3-letter word contains two A's. The letter that is not A can be in the 1st, 2nd, or 3rd position. It can be one of B, C, or D. There are $3 \times 3 = 9$ words for this case.

Case 3: The 3-letter word contains three A's. There's 1 such word, namely AAA.

The total number of words is $27 + 9 + 1 = \boxed{37}$.

2.2.2 There are two cases.

Case 1: The first hat is picked. There are 15 balls in the first hat. The first ball can be any of the 15 balls. The second ball can any of the 14 remaining balls. The third ball can be any of the 13 remaining balls. The number of ordered selections of three balls in this case is $15 \times 14 \times 13 = 2730$.

Case 2: The second hat is picked. There are 10 balls in the second hat. The first ball can be any of the 10 balls. The second ball can any of the 9 remaining balls. The third ball can be any of the 8 remaining balls. The number of ordered selections of three balls in this case is $10 \times 9 \times 8 = 720$.

The total number of possible selections is the sum of the number of selections in each case, which is $2730 + 720 = \boxed{3450}$.

2.2.3 We go through either B or C to get to D. The number of paths going from A to D through B is 2×1, and the number of paths going from A to D through C is 2×3. Thus the total number of paths from A to D is $(2 \times 1) + (2 \times 3) = 8$.

Similarly, to get from D to H, we go through one of E, F, or G. The three cases give a total of $(1 \times 1) + (2 \times 3) + (2 \times 1) = 9$ paths from D to H.

The choice of path from A to D is independent of the choice of path from D to H. Therefore we multiply the number of paths from A to D by the number of paths from D to H to get the number of paths from A to H. The answer is $8 \times 9 = \boxed{72}$.

2.2.4   There are several cases, based on the size of the triangle.

Case 1: The triangle is one of the smallest triangles. There are $1 + 3 + 5 + 7 = 16$ smallest triangles.

Case 2: The triangle is composed of four of the smallest triangles. There are 7 such triangles, 6 pointing up and 1 pointing down.

Case 3: The triangle is composed of nine of the smallest triangles. There are 3 such triangles, all pointing up.

Case 4: The triangle is the biggest triangle. There's 1.

The total number of triangles is $16 + 7 + 3 + 1 = \boxed{27}$.

2.2.5   We count the number of rectangles by cases, based on the side lengths of the rectangle:

Side lengths of rectangle	Number of rectangles
1×1	9
1×2	6
1×3	3
2×1	6
2×2	4
2×3	2
3×1	3
3×2	2
3×3	1

So the number of rectangles whose sides are parallel to the sides of the grid is $9+6+3+6+4+2+3+2+1 = \boxed{36}$.

2.2.6   When we counted the number of rectangles (in the previous problem), we divided the problem into cases based on the side lengths. We use a similar idea when counting isosceles triangles, but this time we divide into cases based on the lengths of the legs and of the base of the triangles.

First we need to find all possible leg lengths. We can begin by looking at all possible distances between two points. Using Pythagorean Theorem, we find the following distances are possible on a 4×4 grid: $1, 2, 3, \sqrt{2}, 2\sqrt{2}, 3\sqrt{2}, \sqrt{5}, \sqrt{10},$ and $\sqrt{13}$.

Now we can count the triangles. To aid our count, note that each triangle is inscribed in a rectangle that we've already counted in the previous problem.

Case #	Leg of triangle	Base of triangle	Triangle inscribed in rectangle	Number of rectangles	Number of triangles in each rectangle	Number of triangles
1	1	$\sqrt{2}$	1×1	9	4	36
2	2	$2\sqrt{2}$	2×2	4	4	16
3	3	$3\sqrt{2}$	3×3	1	4	4
4	$\sqrt{2}$	2	1×2 or 2×1	12	2	24
–	$2\sqrt{2}$	–	–	–	–	0
–	$3\sqrt{2}$	–	–	–	–	0
5	$\sqrt{5}$	$\sqrt{2}$	2×2	4	4	16
6	$\sqrt{5}$	2	2×2	4	4	16
7	$\sqrt{5}$	$\sqrt{10}$	2×3 or 3×2	4	4	16
8	$\sqrt{5}$	$3\sqrt{2}$	3×3	1	4	4
9	$\sqrt{10}$	2	2×3 or 3×2	4	2	8
10	$\sqrt{10}$	$2\sqrt{2}$	3×3	1	4	4
11	$\sqrt{13}$	$\sqrt{2}$	3×3	1	4	4

An example from each case is shown below (the inscribing rectangle is dashed):

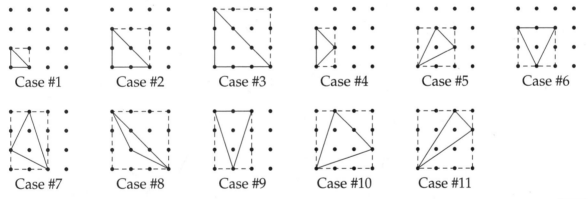

Case #1　　Case #2　　Case #3　　Case #4　　Case #5　　Case #6

Case #7　　Case #8　　Case #9　　Case #10　　Case #11

The total number of isosceles triangles is $36 + 16 + 4 + 24 + 16 + 16 + 16 + 4 + 8 + 4 + 4 = \boxed{148}$.

Exercises for Section 2.3

2.3.1 First we count the number of all 4-letter words with no restrictions on the word. Then we count the number of 4-letter words with no vowels. We then subtract to get the answer.

Each letter of a word must be one of A, B, C, D, or E, so the number of 4-letter words with no restrictions on the word is $5 \times 5 \times 5 \times 5 = 625$. Each letter of a word with no vowel must be one of B, C, or D. So the number of all 4-letter words with no vowels in the word is $3 \times 3 \times 3 \times 3 = 81$. Therefore, the number of 4-letter words with at least one vowel is $625 - 81 = \boxed{544}$.

2.3.2 A 5-digit number can have for its leftmost digit anything from 1 to 9 inclusive, and for each of its next four digits anything from 0 through 9 inclusive. Thus there are $9 \times 10 \times 10 \times 10 \times 10 = 90{,}000$ 5-digit numbers.

A 5-digit number with no zero as a digit can have for each of its five digits anything from 1 through 9 inclusive. There are $9 \times 9 \times 9 \times 9 \times 9 = 59{,}049$ such 5-digit numbers. Therefore the number of 5-digit numbers with at least one zero as a digit is $90{,}000 - 59{,}049 = \boxed{30{,}951}$.

2.3.3 The number of all outfit combinations is $6 \times 6 \times 6 = 216$. There are 6 outfits in which all three items are the same color. Thus there are $216 - 6 = \boxed{210}$ outfits in which not all three items are the same color.

2.3.4 The number of all seating arrangements is $7!$. The number of seating arrangements in which Wilma and Paul sit next to each other is $6! \times 2!$. We can arrive at $6! \times 2!$ by considering Wilma and Paul as one person, arranging the "six" people first, then arranging Wilma and Paul. Thus the number of acceptable arrangements is $7! - 6! \times 2! = \boxed{3600}$.

Exercises for Section 2.4

2.4.1 The first letter can be any of the 26 letters of the alphabet, while the second letter can be any of the 25 remaining letters. The first digit can be any of the 10 digits, while the second digit can be any of the 9 remaining digits. The number of license plates is $26 \times 25 \times 10 \times 9 = \boxed{58{,}500}$.

2.4.2 We proceed using casework on the choice of second digit:

Second digit	First digit
0	1, 2, 3, 4, 5, 6, 7, 8, 9
1	2, 3, 4, 5, 6, 7, 8, 9
2	4, 5, 6, 7, 8, 9
3	6, 7, 8, 9
4	8, 9

The third digit can be any of the 10 digits. The answer is $(9 + 8 + 6 + 4 + 2) \times 10 = \boxed{290}$.

2.4.3

Last digit	First two digits
0	–
1	10
2	11, 20
3	12, 21, 30
4	13, 22, 31, 40
5	14, 23, 32, 41, 50
6	15, 24, 33, 42, 51, 60
7	16, 25, 34, 43, 52, 61, 70
8	17, 26, 35, 44, 53, 62, 71, 80
9	18, 27, 36, 45, 54, 63, 72, 81, 90

The third digit can be any of the 10 digits. The answer is $(1 + 2 + 3 + 4 + 5 + 6 + 7 + 8 + 9) \times 10 = \boxed{450}$.

2.4.4 Regardless of whether x_1 is odd or even, we have 5 choices for x_2: if x_1 is odd then x_2 must be one of the 5 even digits, otherwise if x_1 is even then x_2 must be one of the 5 odd digits. Similarly, we then have 5 choices for x_3, 5 choices for x_4, and so on.

Since x_1 can be any of the 10 digits, the answer is $10 \times 5^5 = \boxed{31{,}250}$.

2.4.5 Let the three numbers be $x < y < z$. We proceed by casework based on the choice of x. For each possible x we list the choices for y such that it is still possible to choose z with $z > xy$.

x	y	z	Number of pairs of (y, z)
$x = 1$	$2 \le y \le 99$	$y < z \le 100$	$98 + 97 + \cdots + 1 = 4851$
$x = 2$	$3 \le y \le 49$	$2y < z \le 100$	$94 + 92 + \cdots + 2 = 2256$
$x = 3$	$4 \le y \le 33$	$3y < z \le 100$	$88 + 85 + \cdots + 1 = 1335$
$x = 4$	$5 \le y \le 24$	$4y < z \le 100$	$80 + 76 + \cdots + 4 = 840$
$x = 5$	$6 \le y \le 19$	$5y < z \le 100$	$70 + 65 + \cdots + 5 = 525$
$x = 6$	$7 \le y \le 16$	$6y < z \le 100$	$58 + 52 + \cdots + 4 = 310$
$x = 7$	$8 \le y \le 14$	$7y < z \le 100$	$44 + 37 + \cdots + 2 = 161$
$x = 8$	$9 \le y \le 12$	$8y < z \le 100$	$28 + 20 + 12 + 4 = 64$
$x = 9$	$10 \le y \le 11$	$9y < z \le 100$	$10 + 1 = 11$

If $10 \le x < y$, then $xy > 100$, so no choice of z is possible.

Therefore, the total number of ways to pick three numbers satisfying the condition of the problem is $4851 + 2256 + 1335 + 840 + 525 + 310 + 161 + 64 + 11 = \boxed{10{,}353}$.

Exercises for Section 2.5

2.5.1 If we consider the group of Democrats as one person, then there are 6! ways to arrange the 6 people (the 5 Republicans and the one Democrat group). Then there are 4! ways to arrange arranging the 4 Democrats within their group. So the number of arrangements is $6! \times 4! = \boxed{17{,}280}$.

2.5.2 First we order the three groups of animals, which we can do in 3! ways. Next we order the animals within each group. There are 4! ways to arrange the group of chickens, 2! ways to arrange the group of dogs, and 5! ways to arrange the group of cats. The answer is $3! \times 4! \times 2! \times 5! = \boxed{34{,}560}$.

2.5.3 *Solution 1*: If the twins sit at the end of the row, then the other sister can sit at one of 5 places so that she's not next to the twins. We first choose one of the 2 ends (2 choices) and seat the 2 twins (2! choices); then we seat the other sister (5 choices); then we seat the remaining 5 brothers (5! choices). We get $2 \times 2! \times 5 \times 5! = 2400$ ways.

If the twins don't sit at the end of the row, the other sister can sit at one of 4 places so that she's not next to the twins. We first pick the twin's seats (5 choices) and arrange the two twins (2! choices); then we seat the other sister (4 choices); then we seat the remaining 5 brothers (5! choices). We get $5 \times 2! \times 4 \times 5! = 4800$ ways.

Adding these two cases, we see that the number of acceptable seatings is $2400 + 4800 = \boxed{7200}$.

Solution 2: We will find the number of ways to seat the twins together, and then subtract the cases where the third sister is sitting next to them. If we treat the twins as a block, then there are 7! ways to arrange the twin block, the other sister, and the 5 boys. However, there are two ways to arrange the twins within their block, so there are $2 \times 7!$ ways to arrange the siblings such that the twins sit together.

If we treat the three sisters as a block, then there are 6! ways to arrange the block and the 5 boys. There are 4 ways to arrange the sisters within the block such that the twins are sitting together (since the non-twin sister cannot sit between her other two sisters), meaning that there are $4 \times 6!$ ways to arrange the siblings such that the twins are together and the three girls are together. So the number of ways to seat the twins together such that the three girls don't sit together is $(2 \times 7!) - (4 \times 6!) = 10080 - 2880 = \boxed{7200}$.

2.5.4

(a) The president can be any one of the 20 members, and the vice-president can be any one of the 19 remaining members. The answer is $20 \times 19 = \boxed{380}$.

(b) The president can be any one of the 20 members, and the vice-president can be any one of the 10 members of the opposite sex. The answer is $20 \times 10 = \boxed{200}$.

(c) The president can be any one of the 20 members, and the vice-president can be any one of the 9 remaining members of the same sex. The answer is $20 \times 9 = \boxed{180}$.

(d) Part (a) can be split into cases (b) and (c), and $380 = 200 + 180$.

2.5.5

(a) The president can be any of the 25 members, the secretary can be any of the 24 remaining members, and the treasurer can be any of the remaining 23 members. There are $25 \times 24 \times 23 = \boxed{13{,}800}$ ways.

(b) There are 25 choices for each position, so there are $25 \times 25 \times 25 = \boxed{15{,}625}$ ways that the positions can be filled.

(c) With no restriction, the president can be any of the 25 members, the secretary can be any of the 25 remaining members, and the treasurer can be any of the remaining 25 members.

If the same member holds all three offices, it could be any of the 25 members, so there are 25 ways for this to happen. We must exclude these 25 possibilities, so the answer is $25 \times 25 \times 25 - 25 = \boxed{15{,}600}$.

Review Problems

2.16 Each of the 4 digits can be one of the 5 odd digits: 1, 3, 5, 7, 9. So there are $5 \times 5 \times 5 \times 5 = \boxed{625}$ such 4-digit numbers.

2.17 The first letter can be one of the 5 vowels, and each of the next two letters can be one of the 26 letters. There are $5 \times 26 \times 26 = \boxed{3380}$ such words.

2.18 We have 5 choices for the first digit, then 4 choices for the second digit (since it must be different from the first), then 4 choices for the third digit (since it must be different from the second). So there are

$5 \times 4 \times 4 = \boxed{80}$ choices for the combination.

2.19 If $a \geq 4$, then $a^3 + b^2 + c > a^3 \geq 4^3 > 50$. But we want $a^3 + b^2 + c \leq 50$, so we must have $a = 2$. Now we substitute $a = 2$ into $a^3 + b^2 + c \leq 50$, which gives $b^2 + c \leq 42$. Since $b^2 < 42$, we know that b must be one of 2, 4, or 6.

When $b = 2$, $c \leq 38$. There are 19 even positive integers less than or equal to 38, namely $2 \times 1, 2 \times 2, \ldots, 2 \times 19$.

When $b = 4$, $c \leq 26$. There are 13 even positive integers less than or equal to 26.

When $b = 6$, $c \leq 6$. There are 3 even positive integers less than or equal to 6.

Thus the answer is $19 + 13 + 3 = \boxed{35}$.

2.20 There are $200 - 100 + 1 = 101$ numbers in the list $100, 101, \ldots, 200$. We can find 5 perfect squares in the list, namely $10^2, \ldots, 14^2$. So the number of non-perfect-squares in the list is $101 - 5 = \boxed{96}$.

2.21 There are only three possibilities for the first digit:

First digit	Last digit
3	1
6	2
9	3

The middle digit can be any of the 10 digits. The answer is $3 \times 10 = \boxed{30}$.

2.22 We can consider the two boys as one person, arrange the "seven" people first, then arrange the 2 boys. So the number of seating arrangements in which the boys sit together is $7! \times 2! = \boxed{10{,}080}$.

2.23 The second digit must be even, so it must be one of 0, 2, 4, 6, or 8. However, it cannot be 6 or 8, since then the fourth digit could not be twice the second digit. Thus, there are 18 different possible combinations of second and fourth digits, as shown in the following table:

Second digit	Fourth digit
0	$0, 1, 2, 3, 4, 5, 6, 7, 8, 9$
2	$4, 5, 6, 7, 8, 9$
4	$8, 9$

The first digit can be any of the 9 nonzero digits, and the third digit can be any of the 10 digits. The answer is $18 \times 9 \times 10 = \boxed{1620}$.

2.24 First we arrange the 2 groups of books; there are $2!$ ways in which we can do this. Then we can arrange the 3 math books in $3!$ ways and the 5 English books in $5!$ ways. Therefore, there are $2! \times 3! \times 5! = \boxed{1440}$ ways to arrange the books.

2.25 If both Penelope and Quentin are not officers, then there are 20 choices for chairman, 19 choices for vice-chairman, and 18 choices for sergeant-at-arms. There are $20 \times 19 \times 18 = 6840$ ways in this case.

If both are officers, Penelope can take one of the 3 positions, Quentin can take one of the 2 remaining positions, and one of the 20 remaining members can take the third position. There are $3 \times 2 \times 20 = 120$ ways in this case.

The answer is $6840 + 120 = \boxed{6960}$.

2.26 The units digit must be 2, 4, 6, or 8 (it can't be 0 because then the number would have to be 000, which isn't allowed). If the units digit is a, then the hundreds digits can be any digit from 1 to a, and the middle digit is then necessarily a minus the hundreds digit. Therefore, there are a such numbers with units digit a. So there are $2 + 4 + 6 + 8 = \boxed{20}$ such numbers.

Challenge Problems

2.27 A naive approach would be to simply list them:

$$25, 34, 43, 52, 59, 61, 68, 70, 77, 86, 95, 106, 115, 124,$$

giving $\boxed{14}$ such numbers. But we might be worried that we skipped one by mistake.

A more systematic approach is to observe that the desired numbers' digital sums could only be either 7 or 14, and furthermore the only ones between 24 and 125 with digital sums of 14 must be 2-digit numbers (since to get a 3-digit number beginning with 1 whose digital sum is 14, the second digit must be at least 4, since the last digit can be at most 9). Therefore, there are 3 exclusive cases:

Case 1: 2-digit numbers whose digital sum is 7. The first digit can be any number a from 2 to 7, and then the second digit must be $7 - a$. So there are 6 of these.

Case 2: 2-digit numbers whose digital sum is 14. The first digit can be any number b from 5 to 9, and then the second digit must be $14 - b$. So there are 5 of these.

Case 3: 3-digit numbers whose digital sum is 7. Since the first digit is 1, then second digit can be any number c from 0 to 2, and then the third number is $6 - c$. So there are 3 of these.

Therefore we add the cases and see that there are $6 + 5 + 3 = \boxed{14}$ such numbers.

2.28 Charlie can take one of the 4 cats, and Danny can take one of the 3 remaining cats, so there are $4 \times 3 = 12$ ways to give cats to these two kids. Since Anna and Betty can't take a goldfish, they select from the 4 remaining animals so there are $4 \times 3 = 12$ ways to give pets to these two kids. For the other three kids, there are $3 \times 2 \times 1 = 6$ ways to give out the remaining 3 pets. The answer is $12 \times 12 \times 6 = \boxed{864}$.

2.29 First, we choose the position that the Grand Pooh-Bah sits in. There are two cases:

Case 1: Person p sits at the far right. There is no further restriction, and the other $n - 1$ members can sit in $(n - 1)!$ ways.

Case 2: Person p sits anywhere other than the far right. There are $n - 1$ choices for where person p sits, and $p - 1$ choices for the person who sits to her immediate right (since it must be one of $1, 2, 3, \ldots, p - 1$). Then the remaining $n - 2$ people can sit in the remaining seats in $(n - 2)!$ ways.

So the total number of seatings is $(n - 1)! + (n - 1)(p - 1)(n - 2)!$, which simplifies to $\boxed{p(n - 1)!}$.

2.30 If we start at one of the 4 N's in a corner, then we have only 1 choice for the first O, then 3 choices for the second O, then 5 choices for the second N, for a total of $4 \times 1 \times 3 \times 5 = 60$ ways to form NOON, starting from a corner. If we start at one of the 8 N's on the side, then we have 2 choices for the first

O, then 3 choices for the second O. If we choose the second O adjacent to our original N, then we only have 4 choices for the final N, otherwise (for the other two choices of the second O) we have 5 choices for the final N. Thus we have $8 \times 2 \times (2 \times 5 + 1 \times 4) = 224$ ways to form NOON, starting from a side. This gives a total of $60 + 224 = \boxed{284}$ ways to form NOON.

2.31

(a) The first two digits determine the palindrome. The first digit can be any of the 9 nonzero digits, and the second digit can be any of the 10 digits. The answer is $9 \times 10 = \boxed{90}$.

(b) The first three digits determine the palindrome. The first digit can be any of the 9 nonzero digits, and each of the second and third digits can be any of the 10 digits. The answer is $9 \times 10 \times 10 = \boxed{900}$.

(c) The first three digits determine the palindrome. The first digit can be any of the 9 nonzero digits, and each of the second and third digits can be any of the 10 digits. The answer is $9 \times 10 \times 10 = \boxed{900}$.

(d) If k is even, then the first $\frac{k}{2}$ digits determine the palindrome. Since there are two choices for each digit, there are $2^{\frac{k}{2}}$ palindromes of length k.

If k is odd, then the first $\frac{k+1}{2}$ digits determine the palindrome. Again, there are two choices for each digit, so there are $2^{\frac{k+1}{2}}$ palindromes of length k. Note that this is the same as the number of palindromes of length $k + 1$.

But, we must also exclude the 2 palindromes that don't have at least one 8 and at least one 9. So we are looking for the smallest odd value of k such that $2^{\frac{k+1}{2}} - 2 \geq 2004$.

We know $2^{10} - 2 = 1022$ and $2^{11} - 2 = 2046$, so we must have $\frac{k+1}{2} \geq 11$, and the smallest such k is $\boxed{21}$.

2.32   There are three types of planes that can go through three points in the cube. The plane can contain an entire face of the cube, or it can contain parallel opposite edges of the cube (not on the same face), or it can fail to contain an edge of the cube.

Case 1: The plane contains a face. There are 6 faces, and therefore 6 planes that pass through them.

Case 2: The plane contains parallel opposite edges. There are 12 edges, and so $12/2 = 6$ pairs of parallel opposite edges.

Case 3: The plane does not contain an edge. For example, in the picture to the right, the plane passes through vertices A, B, and C. Note that A, B and C all share an edge with vertex D. Every plane which does not contain an edge of the cube must pass through 3 vertices which all share an edge with a unique vertex as in this example, and each vertex uniquely determines such a plane. Since the cube has 8 vertices, there are 8 planes that do not contain an edge of the cube.

The sum of the three cases is $6 + 6 + 8 = \boxed{20}$ planes.

2.33   The numbers 1 to 255 are all the k-digit binary numbers with $k = 1, 2, \ldots, 8$.

First, let's look at the 8-digit binary numbers $\overline{1a_6a_5a_4a_3a_2a_1a_0}$. In a number with $a_0 = 0$, each of the 6 digits $a_1, a_2, \ldots, a_6$ can be 0 or 1. Thus the number of 8-digit binary numbers with $a_0 = 0$ is $2^6 = 64$.

Similarly, the number of 8-digit binary numbers with $a_k = 0$ is 64 for $k = 1, 2, \ldots, 6$. Therefore the total number of 0's in all 8-digit binary numbers is 64×7.

Similarly we can show the following:

number of digits	Total number of zeros
8	64×7
7	32×6
6	16×5
5	8×4
4	4×3
3	2×2
2	1×1
1	0×0

Finally, the number 256 has 8 zeros in binary. Therefore, the answer is $(64 \times 7) + (32 \times 6) + (16 \times 5) + (8 \times 4) + (4 \times 3) + (2 \times 2) + (1 \times 1) + (0 \times 0) + 8 = \boxed{777}$.

2.34   The total number of all 5-letter words is 26^5, since each letter can be any letter of the alphabet. The number of 5-letter words in which no two consecutive letters are the same is 26×25^4. This is because the first letter can be anything, while we have 25 choices for each subsequent letter, since each letter cannot match the previous one. We then subtract to get our answer $26^5 - (26 \times 25^4) = \boxed{1{,}725{,}126}$.

2.35   Since we're looking at relative positions, it doesn't matter where the first student sits. The second student can sit left or right of the first. The third student can sit left or right of the block of the two already seated students. The fourth student can sit left of right of the block of three already seated students, and so on. Finally, the ninth student can sit left or right of the block of eight. The tenth student takes the only open seat left. Each of students 2 through 9 has 2 choices, so the number of seatings is $2^8 = \boxed{256}$.

2.36   We could proceed by listing the various cases, depending on which number Mathew draws.

Mathew's number	My pair of numbers
1	−
2	−
3	$(1,2), (2,1)$
4	$(1,3), (3,1)$
5	$(1,4), (2,3), (3,2), (4,1)$
6	$(1,5), (2,4), (4,2), (5,1)$
7	$(1,6), (2,5), (3,4), (4,3), (5,2), (6,1)$
8	$(2,6), (3,5), (5,3), (6,2)$
9	$(3,6), (4,5), (5,4), (6,3)$
10	$(4,6), (6,4)$
11	$(5,6), (6,5)$
12	−

The answer is $2 + 2 + 4 + 4 + 6 + 4 + 4 + 2 + 2 = \boxed{30}$.

There is a much easier solution: there are $6 \times 5 = 30$ choices for the two marbles that I draw. Once I draw my two marbles, there is only 1 way for Mathew to draw the marble which has the sum of my two marbles. So the total number of possibilities is just equal to the number of ways in which I can draw my two marbles, which is $\boxed{30}$.

CHAPTER **3**

_____ **Correcting for Overcounting**

Exercises for Section 3.2

3.2.1

(a) | DEEG, DEGE, DGEE, EDEG, EDGE, EEDG, EEGD, EGDE, EGED, GDEE, GEDE, GEED |

(b) This is counting the number of ways four distinct objects can be put in order, so there are 4! = $\boxed{24}$ different arrangements.

(c) If the E's are not unique, then each arrangement from part (a) gets counted twice in part (b): for example, EDGE from part (a) is counted as both E_1DGE_2 and E_2DGE_1 in part (b). So the 24 arrangements in part (b), divided by the 2! ways to arrange two E's, gives the 12 arrangements in part (a).

3.2.2

(a) This is counting the number of ways that six distinct objects can be put in order, so there are 6! = $\boxed{720}$ different arrangements.

(b) For any given arrangement from part (a), there are 3! ways to arrange the three A's within the word, and there are 2! ways to arrange the two N's. So we divide these two numbers into the number of arrangements if the letters were different (that we found in part (a)) and we find the answer is $\dfrac{720}{2! \times 3!} = \boxed{60}$.

3.2.3

(a) All the letters are unique so we find the number of permutations for 3 objects. 3! = $\boxed{6}$.

(b) All the letters are unique so we find the number of permutations for 3 objects. 3! = $\boxed{6}$.

(c) First we count the arrangements if the two P's are unique, which is 3!. Then since the P's are not unique, we divide by 2! for the arrangements of P, for an answer of $\dfrac{3!}{2!} = \boxed{3}$.

(d) First we count the arrangements if the two T's are unique, which is 4!. Then since the T's are not unique, we divide by 2! for the arrangements of T, for an answer of $\dfrac{4!}{2!} = \boxed{12}$.

(e) First we count the arrangements if the three L's are unique, which is 4!. Then since the L's are not unique, we divide by 3! for the arrangements of L, for an answer of $\frac{4!}{3!} = \boxed{4}$.

(f) First we count the arrangements if the two E's are unique, which is 5!. Then since the E's are not unique, we divide by 2! for the arrangements of E, for an answer of $\frac{5!}{2!} = \boxed{60}$.

(g) First we count the arrangements if all the letters are unique, which is 4!. Then since the T's and the O's are not unique, we divide by 2! twice for the arrangements of T's and the arrangement of O's, for an answer of $\frac{4!}{2! \times 2!} = \boxed{6}$.

(h) First we count the arrangements if the three E's are unique, which is 5!. Then since the E's are not unique, we divide by 3! for the arrangements of E, for an answer of $\frac{5!}{3!} = \boxed{20}$.

(i) First we count the arrangements if all the letters are unique, which is 5!. Then since the M's and the A's are not unique, we divide by 2! twice for the arrangements of M's and the arrangements of A's, for an answer of $\frac{5!}{2! \times 2!} = \boxed{30}$.

(j) First we count the arrangements if all the letters are unique, which is 6!. Then since the T's, A's, and the R's are not unique, we divide by 2! thrice for the arrangements of T's, A's, and R's, for an answer of $\frac{6!}{2! \times 2! \times 2!} = \boxed{90}$.

(k) First we count the arrangements if the four A's are unique, which is 7!. Then since the A's are not unique, we divide by 4! for the arrangements of A, for an answer of $\frac{7!}{4!} = \boxed{210}$.

(l) First we count the arrangements if all the letters are unique, which is 11!. Then since the I's, S's and the P's are not unique, we divide by 4!, 4!, and 2! for the arrangements of I's, S's, and P's, for an answer of $\frac{11!}{4! \times 4! \times 2!} = \boxed{34{,}650}$.

3.2.4 There are 5! ways to arrange the books if they are unique, but two are identical so we must divide by 2! for an answer of $\frac{5!}{2!} = \boxed{60}$.

3.2.5 There are 8! ways to arrange 8 allocations to the pens if they are not identical, but we must divide by 4! for the four dog pen allocations and divide by 3! for the three cat pen allocations. So the answer is $\frac{8!}{4! \times 3!} = \boxed{280}$.

Exercises for Section 3.3

3.3.1 If the co-president positions are unique, there are 15 choices for the first president and 14 choices for the second president. However, since the positions are identical, we must divide by 2!, or the number of arrangements of the co-president positions, to get $\frac{15 \times 14}{2!} = \boxed{105}$ ways.

3.3.2 If the order the balls are selected matters, there are 20 choices for the first selection and 19 choices for the second selection. However, since the ball drawings are identical, we must divide by 2!, which is the number of arrangements of each pair of balls, to get $\frac{20 \times 19}{2!} = \boxed{190}$.

3.3.3 Each team plays 6 other teams in its division twice, and the 7 teams in the other division once, for a total of $6 \times 2 + 7 = 19$ games for each team. There are 14 teams total, which gives a preliminary count of $19 \times 14 = 266$ games, but we must divide by two because we have counted each game twice (once for one team and once for the other). So the final answer is $\dfrac{19 \times 14}{2} = \boxed{133}$ games.

3.3.4

(a) Let $S = 2 + 4 + 6 + \cdots + 2n$. Note that S has n terms. Another expression for S is $2n + \cdots + 6 + 4 + 2$, so adding the first and second expressions for S, we obtain the equation $2S = (2n + 2) + (2n + 2) + \cdots + (2n + 2)$. This sum has n terms, so $2S = n(2n + 2) = 2n(n + 1)$, and $\boxed{S = n(n + 1)}$.

Alternatively, note that $S = 2(1 + 2 + 3 + \cdots + n) = 2(\frac{n(n+1)}{2}) = n(n + 1)$.

(b) Let $S = 1 + 3 + 5 + \cdots + (2n - 1)$. Note that S has n terms. Another expression for S is $(2n - 1) + \cdots + 5 + 3 + 1$, so adding the first and second expressions for S, we obtain the equation $2S = 2n + 2n + \cdots + 2n$. This sum has n terms, so $2S = n(2n) = 2n^2$, and $S = \boxed{n^2}$.

Also note that $S = (2 - 1) + (4 - 1) + (6 - 1) + \cdots + (2n - 1) = (2 + 4 + 6 + \cdots + 2n) - (n \cdot 1) = n(n + 1) - n = n^2$.

3.3.5 There are 12 vertices in the icosahedron, so from each vertex there are potentially 11 other vertices to which we could extend a diagonal. However, 5 of these 11 points are connected to the original point by an edge, so they are not connected by interior diagonals. So each vertex is connected to 6 other points by interior diagonals. This gives a preliminary count of $12 \times 6 = 72$ interior diagonals. However, we have counted each diagonal twice (once for each of its endpoints), so we must divide by 2 to correct for this overcounting, and the answer is $\dfrac{12 \times 6}{2} = \boxed{36}$ diagonals.

Exercises for Section 3.4

3.4.1 There are 8! ways to place the people around the table, but this counts each valid arrangement 8 times (once for each rotation of the same arrangement). The answer is $\dfrac{8!}{8} = 7! = \boxed{5040}$.

3.4.2 There are 5! ways to place the keys on the keychain, but we must divide by 5 for rotational symmetry (5 rotations for each arrangement), and by 2 for reflectional symmetry (we can flip the keychain to get the same arrangement). The answer is $\dfrac{5!}{5 \times 2} = \boxed{12}$.

3.4.3

(a) There are 10 people to place, so we can place them in 10! ways, but this counts each valid arrangement 10 times (once for each rotation of the same arrangement). So the number of ways to seat them is $\dfrac{10!}{10} = 9! = \boxed{362{,}880}$.

(b) Choose any 5 consecutive seats in which to place the Democrats—it doesn't matter which 5 consecutive seats that we choose, since we can rotate the table. Then there are 5! ways to place the Democrats in their seats, and 5! ways to place the Republicans in their seats, for a total of $5! \times 5! = \boxed{14{,}400}$ arrangements.

(c) The only way that the Senators can be seated is if the seats alternate by party, as shown to the right (where D is a Democratic seat and R is a Republican seat). Fix the rotation by placing the youngest Democrat in the top seat, so that we have removed the overcounting of rotations of the same arrangement. Now there are 4! ways to place the remaining Democrats in the other Democratic seats, and 5! ways to place the Republicans in the Republican seats, for a total of $5! \times 4! = \boxed{2{,}880}$ arrangements.

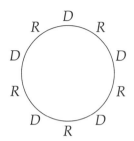

3.4.4 There are 6 choices of seats for Fred to sit in. Once Fred is seated, then Gwen must sit opposite him. This leaves 4 people to place in the four remaining seats, which can be done in 4! ways. However, we must divide by 6 to account for the 6 rotations of the table. So the number of arrangements is $\frac{6 \times 1 \times 4!}{6} = 4! = \boxed{24}$.

Alternatively, we could start by fixing the table around Fred, thus removing the rotation. There is 1 option for Gwen's seat, since she must sit across from him. This leaves 4 people to place in four unique seats, so the number of arrangements is $4! = \boxed{24}$.

3.4.5 There are 8 choices for seats for Alice to sit in. Once Alice is seated, there are 5 seats left for Bob, since he won't sit in either seat immediately next to Alice. This leaves 6 people to place in the remaining 6 seats, which can be done in 6! ways. However, we must divide by 8 to account for the 8 rotations of the table. So the number of arrangements is $\frac{8 \times 5 \times 6!}{8} = 5 \times 6! = \boxed{3600}$.

Alternatively, we can account for the rotations at the beginning, by fixing the table around Alice. Bob can't sit in her seat, or the two seats next to her. This leaves 5 places for him to sit. Then, this leaves 6 unique seats for the 6 remaining people, so there are 6! ways to sit them after Bob is sitting. So the answer is $5 \times 6! = \boxed{3600}$.

Review Problems

3.11

(a) This is the number of ways to arrange 5 unique objects, which is $5! = \boxed{120}$.

(b) $ste_1e_2e_3, ste_1e_3e_2, ste_2e_1e_3, ste_2e_3e_1, ste_3e_1e_2, ste_3e_2e_1$, giving $\boxed{6}$ arrangements.

(c) steee, setee, seete, seeet, estee, esete, eseet, eeste, eeset, eeest, tseee, tesee, teese, teees, etsee, etese, etees, eetse, eetes, eeets, giving $\boxed{20}$ arrangements.

(d) Yes, we see that $\frac{120}{6} = 20$. The number q counts the number of ways that each arrangement from part (c) is overcounted in part (a), so we must divide p by q to get the number of arrangements r in part (c).

3.12

(a) All the letters are unique so we find the number of permutations for 3 objects. $3! = \boxed{6}$.

(b) First we count the arrangements if the two D's are unique, which is 3!. Then since the D's are not unique, we divide by 2! for the arrangements of the D's, for an answer of $\frac{3!}{2!} = \boxed{3}$.

(c) All the letters are unique so we find the number of permutations for 4 objects. $4! = \boxed{24}$.

(d) First we count the arrangements if the two N's are unique, which is 4!. Then since the N's are not unique, we divide by 2! for the arrangements of the N's, for an answer of $\frac{4!}{2!} = \boxed{12}$.

(e) First we count the arrangements if the two I's are unique, which is 5!. Then since the I's are not unique, we divide by 2! for the arrangements of the I's, for an answer of $\frac{5!}{2!} = \boxed{60}$.

(f) There are two O's and five total letters, so the answer is $\frac{5!}{2!} = \boxed{60}$.

(g) There are two C's and six total letters, so the answer is $\frac{6!}{2!} = \boxed{360}$.

(h) There are three E's and six total letters, so the answer is $\frac{6!}{3!} = \boxed{120}$.

(i) There are two A's, two M's, and six total letters, so the answer is $\frac{6!}{2! \times 2!} = \boxed{180}$.

(j) There are two E's, two L's, and seven total letters, so the answer is $\frac{7!}{2! \times 2!} = \boxed{1260}$.

(k) There are three A's and eight total letters, so the answer is $\frac{8!}{3!} = \boxed{6720}$.

(l) There are two O's, two I's, two N's and eleven total letters, so the answer is $\frac{11!}{2! \times 2! \times 2!} = \boxed{4,989,600}$.

3.13 There are 8! ways to arrange the books if they are unique, but we must divide out the number of permutations for the identical math books (3!), identical English books (3!), and the identical French books (2!). This gives a final answer of $\frac{8!}{3! \times 3! \times 2!} = \boxed{560}$ arrangements.

3.14 First we place the 0, which we only have four options for (everywhere but the first digit). Then we have 4 remaining places to put the last 4 digits, two of which are not unique (the fives), so there are $\frac{4!}{2!}$ options for arranging the other 4 digits. This gives a final answer of $\frac{4 \times 4!}{2!} = \boxed{48}$.

3.15 There are 5! ways to arrange 5 beads in a line. Since there are 5 rotations in a circle for each of these arrangements, we must divide by 5, and since there are two matching reflections for each arrangement, we must divide by 2. So there are $\frac{5!}{5 \times 2} = \boxed{12}$ ways.

3.16 There are 9! ways to arrange 9 people in a line, however there are 9 identical rotations for each arrangement, so we divide by 9 to get $\frac{9!}{9} = 8! = \boxed{40,320}$.

3.17 First choose three consecutive seats for Pierre, Rosa, and Thomas. It doesn't matter which three consecutive seats that we choose, since any three such seats can be rotated to any other such seats. Once the three seats are chosen, there are 3! ways to seat the three friends there. The other five seats are for the other five people, so there are 5! ways to seat them there. The answer is $3! \times 5! = \boxed{720}$.

3.18 All 12 people shake hands with 10 other people (everyone except themselves and their spouse). In multiplying 12×10, each handshake is counted twice, so we divide by two to get the answer of $\frac{12 \times 10}{2} = \boxed{60}$ handshakes.

3.19 We can choose two out of seven points (without regard to order) in $\frac{7 \times 6}{2} = 21$ ways, so there are $\boxed{21}$ chords.

3.20 Each pair of lines will give two pairs of vertical angles. There are $\frac{5 \times 4}{2} = 10$ ways to choose a pair of lines, therefore there are $2 \times 10 = \boxed{20}$ pairs of vertical angles.

Challenge Problems

3.21

(a) Fix the seating around Aubrey. There is only 1 option for where Fred can sit, which is directly across from Aubrey. There are 8! arrangements to place the 8 kids in the other seats, so the number of arrangements is $8! = \boxed{40{,}320}$.

(b) As in part (a), fix the seating around Aubrey. There is only 1 option for Fred to sit, which is directly across from Aubrey. There are 8 remaining places for Betty sit, and once she sits, Guang's location is determined, so there is only 1 option for him. This leaves 6 people and 6 seats remaining, and there are 6! ways to place them. So there are $8 \times 6! = \boxed{5760}$ arrangements.

3.22 There are 8 ways to select the child in the middle, and then we must arrange the remaining 7 children in a circle. There are 7! ways to arrange the 7 remaining children in a line, so when we arrange them in a circle, we must divide by 7 to correct for the rotation. So the answer is $\frac{8 \times 7!}{7} = \boxed{5760}$.

3.23 Each team plays the four other teams within their division three times, for a total of 12 intra-division games. Each team plays the other 15 teams twice for a total of 30 inter-division games, for a total of 42 games per team. However, we have counted every game twice (once for either team) so we must divide by two. The answer is $\frac{20 \times 42}{2} = \boxed{420}$.

3.24 There are three diagonals from any point not parallel to any side, and there are 8 points. We must divide by two because each diagonal was counted twice (once from either of its endpoints). So there are $\frac{8 \times 3}{2} = \boxed{12}$ such diagonals.

3.25 We can break up the sum into a difference of two sums that we already know the formula for:

$$j + (j + 1) + \cdots + (k - 1) + k = (1 + 2 + \cdots + (k - 1) + k) - (1 + 2 + \cdots + (j - 1)).$$

So the sum is equal to

$$\frac{k(k + 1)}{2} - \frac{(j - 1)j}{2} = \boxed{\frac{k(k + 1) - j(j - 1)}{2}}.$$

Alternatively, note that

$$j + (j + 1) + \cdots + (k - 1) + k = (1 + (j - 1)) + (2 + (j - 1)) + \cdots + ((k - j) + (j - 1)) + ((k - j + 1) + (j - 1)).$$

So we can factor out all of the extra $(j - 1)$ terms, and the sum is equal to

$$(1 + 2 + \cdots + (k - j) + (k - j + 1)) + (k - j + 1)(j - 1).$$

This then gives

$$\frac{(k - j + 1)(k - j + 2)}{2} + (k - j + 1)(j - 1).$$

You can check that this is the same expression as our first solution.

3.26 Each face must have a different color. Fix the red face. The other three faces share one edge with the red face, so the red face shares an edge with the blue face, the green face, and the orange face. We can fix the position of the blue face; there are then $2! = \boxed{2}$ ways to color the other two faces.

Alternatively, we could initially color the faces in $4! = 24$ ways. Then we note that there are 12 ways to rotate a tetrahedron (try to figure this out on your own), so we must divide our initial count by 12 to account for these symmetries, giving $\frac{4!}{12} = \boxed{2}$ ways to color the tetrahedron.

3.27 First, fix the position of the first house key, which eliminates the rotations of the keyring. Then place the second house key next to the first one—it doesn't matter on which side of the first one we place it, since we can flip the keyring to get it from one side to the other. So we have accounted for all of the symmetry of the keyring, and we must place the remaining 5 keys. If we think of the 2 car keys as a block, we can place the car-key-block and the 3 office keys in 4! ways. Then, there are 2! ways to arrange the 2 car keys within the block. This gives an answer of $2! \times 4! = \boxed{48}$ ways to arrange the keys.

3.28 There are 8! ways to seat the 8 people if we ignore the symmetry. However, there are 4 different ways in which we can rotate the table: we can rotate it 90 degrees clockwise, 90 degrees counterclockwise, 180 degrees, or not at all. So we must divide our count by 4 to get the answer: $\frac{8!}{4} = \boxed{10{,}080}$.

Alternatively, if we wish to fix the rotation by placing the first person, we have 2 choices for how to place her: either she sits in a seat with a corner to her immediate left or she sits in a seat with a corner to her immediate right. We then have 7! ways to place to remaining 7 people. So the total number of seatings is $2 \times 7! = \boxed{10{,}080}$.

CHAPTER 4

Committees and Combinations

Exercises for Section 4.2

4.2.1

(a) Recall that when choosing officers, order matters. The first position can be any of the 9 people. The second position can be any of the remaining 8 people, and so on. The answer is $9 \times 8 \times 7 \times 6 = \boxed{3024}$.

(b) Recall that choosing a committee is a combination, and order does not matter. We are choosing a 4-person committee from 9 people, so there are $9 \times 8 \times 7 \times 6$ ways to pick the four people for the positions, but then we must divide by 4! since order doesn't matter, so the answer is $\dfrac{9 \times 8 \times 7 \times 6}{4!} = \boxed{126}$.

4.2.2 Choosing a committee is a combination, since the order does not matter. We are choosing a 4-person committee from 25 people, so there are 25 ways to pick the first person, 24 ways to pick the second person, etc. However, we must divide by 4! since order doesn't matter. So the answer is $\dfrac{25 \times 24 \times 23 \times 22}{4!} = \boxed{12{,}650}$.

4.2.3 First we choose the goalie, and any of the 15 people can be the goalie. Then we choose 6 more players from the remaining 14 players, which is same as choosing a committee. There are 14 ways to choose the first player, 13 ways to choose the second player, and so on, down to 9 ways to choose the sixth player. We must then divide by 6! since order of the six players doesn't matter. So the answer is $\dfrac{15 \times 14 \times 13 \times 12 \times 11 \times 10 \times 9}{6!} = \boxed{45{,}045}$.

4.2.4 No three vertices are collinear, so any combination of 3 vertices will make a triangle. There are 8 ways to choose the first point, 7 ways to choose the second point, and 6 ways to choose the third point, but we must divide by 3! since order doesn't matter. So the answer is $\dfrac{8 \times 7 \times 6}{3!} = \boxed{56}$.

4.2.5 There are 55 ways to choose the first Republican, 54 ways to choose the second Republican, and 53 ways to choose the third Republican, however we must divide by 3! because order does not matter. So the number of ways to choose Republicans is $\dfrac{55 \times 54 \times 53}{3!} = 26{,}235$. There are 45 ways to choose the first Democrat and 44 ways to choose the second Democrat, but we must divide by 2! because order does not matter. So the number of ways to choose the Democrats is $\dfrac{45 \times 44}{2!} = 990$. So there are

$26{,}235 \times 990 = \boxed{25{,}972{,}650}$ ways to choose a committee.

4.2.6   There are 30 ways to draw the first white ball, 29 ways to draw the second, and 28 ways to draw the third. However, since order doesn't matter, we must divide by 3! to get $\dfrac{30 \times 29 \times 28}{3!} = 4060$ ways to draw three white balls. There are 20 ways to draw the first red ball and 19 ways to draw the second, however, since order doesn't matter, we must divide by 2! to get $\dfrac{20 \times 19}{2!} = 190$ ways to draw two red balls. So the total number of outcomes for both the red and the white balls is $4060 \times 190 = \boxed{771{,}400}$.

Exercises for Section 4.3

4.3.1

(a) $\dbinom{5}{1} = \dfrac{5!}{1!4!} = \dfrac{(5 \times 4 \times 3 \times 2)(1)}{(1)(4 \times 3 \times 2 \times 1)} = \dfrac{5}{1} = \boxed{5}$.

(b) $\dbinom{5}{3} = \dfrac{5!}{3!2!} = \dfrac{(5 \times 4)(3 \times 2 \times 1)}{(3 \times 2 \times 1)(2 \times 1)} = \dfrac{5 \times 4 \times 3}{3 \times 2 \times 1} = \boxed{10}$.

(c) $\dbinom{7}{4} = \dfrac{7!}{4!3!} = \dfrac{7 \times 6 \times 5 \times 4}{4 \times 3 \times 2 \times 1} = \boxed{35}$.

(d) $\dbinom{8}{2} = \dfrac{8!}{2!6!} = \dfrac{8 \times 7}{2 \times 1} = \boxed{28}$.

(e) $\dbinom{9}{8} = \dfrac{9!}{8!1!} = \dfrac{9 \times 8 \times 7 \times 6 \times 5 \times 4 \times 3 \times 2}{8 \times 7 \times 6 \times 5 \times 4 \times 3 \times 2 \times 1} = \boxed{9}$.

(f) $\dbinom{10}{4} = \dfrac{10!}{4!6!} = \dfrac{10 \times 9 \times 8 \times 7}{4 \times 3 \times 2 \times 1} = \boxed{210}$.

(g) $\dbinom{11}{9} = \dfrac{11!}{9!2!} = \dfrac{11 \times 10 \times 9 \times 8 \times 7 \times 6 \times 5 \times 4 \times 3}{9 \times 8 \times 7 \times 6 \times 5 \times 4 \times 3 \times 2 \times 1} = \boxed{55}$.

(h) $\dbinom{50}{2} = \dfrac{50!}{2!48!} = \dfrac{50 \times 49}{2 \times 1} = \boxed{1225}$.

(i) $\dbinom{1293}{1} = \dfrac{1293!}{1!1292!} = \dfrac{1293}{1} = \boxed{1293}$.

4.3.2   Recall that by definition $0! = 1$. Thus, $\dbinom{n}{0} = \dfrac{n!}{0!n!} = \boxed{1}$. Also, the only way to choose 0 objects out of n is not to choose any of them, so $\binom{n}{0} = 1$.

4.3.3 $\dbinom{n}{1} = \dfrac{n!}{1!(n-1)!} = \boxed{n}$. Also, $\binom{n}{1}$ is the number of ways to choose 1 object out of n. Since there are n different objects, there are n ways to do this.

4.3.4 $\binom{n}{n} = \frac{n!}{n!0!} = \boxed{1}$. Also, there is only one way to choose n objects out of n, which is simply choosing all of them.

4.3.5 By the definition of combination,

$$\binom{5}{7} = \frac{5!}{7!(-2)!},$$

but $(-2)!$ is undefined. So it seems to not make sense. On the other hand, we can write

$$\binom{5}{7} = \frac{5 \cdot 4 \cdot 3 \cdot 2 \cdot 1 \cdot 0 \cdot (-1)}{7!},$$

which is 0 since there's a 0 in the numerator. Finally, we can think of $\binom{5}{7}$ as the number of ways to choose 7 objects from a set of 5 objects. Of course, this is impossible, so there are 0 ways to choose these objects. So $\binom{5}{7} = \boxed{0}$.

Exercises for Section 4.4

4.4.1

(a) Both are equal to $\binom{3}{1} = \frac{3!}{2!1!} = \frac{3}{1} = \boxed{3}$.

(b) Both are equal to $\binom{7}{2} = \frac{7!}{5!2!} = \frac{7 \times 6}{2 \times 1} = \boxed{21}$.

(c) Both are equal to $\binom{11}{3} = \frac{11!}{8!3!} = \frac{11 \times 10 \times 9}{3 \times 2 \times 1} = \boxed{165}$.

(d) Both are equal to $\binom{68}{2} = \frac{68!}{66!2!} = \frac{68 \times 67}{2 \times 1} = \boxed{2278}$.

4.4.2

(a) $\binom{11}{9} = \binom{11}{2} = \frac{11 \times 10}{2!} = \boxed{55}$.

(b) $\binom{16}{15} = \binom{16}{1} = \boxed{16}$. See 4.3.3.

(c) $\binom{30}{27} = \binom{30}{3} = \frac{30 \times 29 \times 28}{3!} = \boxed{4060}$.

(d) $\binom{182}{180} = \binom{182}{2} = \frac{182 \times 181}{2!} = \boxed{16{,}471}$.

(e) $\binom{505}{505} = \binom{505}{0} = \boxed{1}$. See 4.3.2.

4.4.3 First we choose the goalie, and any of the 16 people can be the goalie. Then we choose 10 more players from the remaining 15 players, which is same as choosing a committee. The answer is

$$16\binom{15}{10} = 16\binom{15}{5} = 16 \times \frac{15 \times 14 \times 13 \times 12 \times 11}{5 \times 4 \times 3 \times 2 \times 1} = \boxed{48,048}.$$

4.4.4 $\binom{n}{n-1} = \binom{n}{1} = \boxed{n}$. See 4.3.3.

Review Problems

4.11.

(a) $\binom{7}{2} = \frac{7 \times 6}{2} = \boxed{21}$.

(b) $\binom{8}{6} = \binom{8}{2} = \frac{8 \times 7}{2} = \boxed{28}$.

(c) $\binom{12}{9} = \binom{12}{3} = \frac{12 \times 11 \times 10}{3 \times 2 \times 1} = \boxed{220}$.

(d) $\binom{16}{5} = \frac{16 \times 15 \times 14 \times 13 \times 12}{5 \times 4 \times 3 \times 2 \times 1} = \boxed{4368}$.

(e) $\binom{85}{82} = \binom{85}{3} = \frac{85 \times 84 \times 83}{3 \times 2 \times 1} = \boxed{98,770}$.

(f) $\binom{133}{133} = \binom{133}{0} = \boxed{1}$.

4.12 Since we don't care about the order that the books are chosen, we have $\binom{7}{3} = \boxed{35}$ possible choices for 3 books out of 7.

4.13 Order does not matter, so it is a combination. Choosing 3 out of 8 is $\binom{8}{3} = \boxed{56}$.

4.14 There are $\binom{5}{2} = 10$ ways to choose the first two buttons, and then 5 choices for the last button. So there are $10 \times 5 = \boxed{50}$ possible combinations.

4.15 Order does not matter, so it is a combination. Choosing 4 from 12 is $\binom{12}{4} = \boxed{495}$.

4.16 No three vertices are collinear, so any combination of 3 vertices will make a triangle. Choosing 3 out of 12 is $\binom{12}{3} = \boxed{220}$.

4.17 The 4×4 grid of points forms 4 horizontal lines and 4 vertical lines. Any choice of two distinct horizontal lines and two distinct vertical lines determines a rectangle whose sides are parallel to the sides of the grid. The number of ways to choose 2 lines out of 4 is $\binom{4}{2} = 6$. Hence the number of rectangles whose sides are parallel to the sides of the grid is $6 \cdot 6 = \boxed{36}$.

Challenge Problems

4.18

(a) Recall from 4.3.2 and 4.3.3 that for any n, $\binom{n}{0} = 1$ and $\binom{n}{1} = n$. Thus

$$\binom{4}{0} + \binom{4}{1} + \binom{4}{2} + \binom{4}{3} + \binom{4}{4} = 1 + 4 + \frac{4 \times 3}{2} + 4 + 1 = \boxed{16}.$$

(b) $\binom{6}{1} + \binom{6}{2} = 6 + \frac{6 \times 5}{2} = \boxed{21}$.

(c) $\binom{7}{2} = \frac{7 \times 6}{2} = \boxed{21}$.

(d) Consider a 2-person committee chosen from a group of 7 people. We know the number of committees is given by part (c). On the other hand, let Adam be one of the 7 people. If Adam is on the committee, we choose 1 more committee member from the remaining 6. There are $\binom{6}{1}$ ways. If Adam is not on the committee, we need to choose 2 from the remaining 6. There are $\binom{6}{2}$ ways. These two cases are the only ways to form a 2-person committee from 7 people: Adam is either on or not on the committee. So parts (b) and (c) each count the number of 2-person committees that can be formed from a group of 7 people, and thus are equal.

4.19 Here we use casework, based on the number of toppings that Sarah chooses. There are a total of 5 toppings, and her number of toppings can be 0, 1, or 2. The order of the toppings does not matter, so each case is counted by a combination. Thus the answer is $\binom{5}{0} + \binom{5}{1} + \binom{5}{2} = \boxed{16}$.

4.20 We approach this problem using complementary counting. If we ignore the restriction regarding Rocco, then the total number of committees is $\binom{12}{3}\binom{12}{4}$, since there are $\binom{12}{3}$ ways to choose 3 people (out of 12) for the Helicopter Committee and $\binom{12}{4}$ ways to choose 3 people (out of 12) for the Glider Committee, and the choices of the two committees are independent. Now we subtract the count of what we don't want. The total number of committees with Rocco serving on both committees is $\binom{11}{2}\binom{11}{3}$, because we choose the 2 other members for the Helicopter Committee (from the 11 remaining club members)and choose the 3 other members for the Glider Committee (again from the 11 remaining club members). Therefore the final answer is $\binom{12}{3}\binom{12}{4} - \binom{11}{2}\binom{11}{3} = \boxed{99,825}$.

4.21

(a) The order of the states does not matter, so this is combination. Choosing 3 states out of 50 is $\binom{50}{3} = \boxed{19,600}$.

(b) We want to choose 1 of the 2 Senators. The answer is $\binom{2}{1} = \boxed{2}$.

(c) First we choose 3 states from 50, then for each state we choose 1 of the 2 Senators. Thus the answer is $19,600 \times 2^3 = \boxed{156,800}$.

4.22

(a) There are 5 pairs and we need to select 2 without regard to order, so there are $\binom{5}{2} = \boxed{10}$ choices.

(b) It is easier to count the ways to fail—namely, the number of ways to select 4 gloves so that we don't have a pair. We have to take one glove from each of 4 different pairs. There are $\binom{5}{4} = 5$

choices for the 4 pairs of gloves, then 2 choices for the glove within each pair (left or right), for a total of $5 \times 2^4 = 80$ choices. Without caring about the restriction, there are $\binom{10}{4} = 210$ ways to choose any 4 gloves. Therefore, the number of ways to choose 4 gloves such that we get at least one pair is $210 - 80 = \boxed{130}$.

4.23 The three cards of same rank can be any of the 13 ranks, and the suits have $\binom{4}{3}$ possible combinations. The next card can be any of the 12 remaining ranks and any of the 4 suits. The last card can be any of the 11 remaining ranks and any of the 4 suits. However, the last two cards are interchangeable (their order doesn't matter), so we must divide by 2! to correct for overcounting. Thus the number of 3-of-a-kind hands is $\dfrac{(13 \times 4) \times (12 \times 4) \times (11 \times 4)}{2!} = \boxed{54{,}912}$.

4.24 Let's look at where the N's will be. Once the positions of the N's are determined, the arrangement of the word is determined. We pick 4 positions for the N's out of 9. Since the N's are not distinguishable from each other, this is a combination. The answer is $\binom{9}{4} = \boxed{126}$.

4.25

(a) There are 8 vertices. The order of endpoints for a segment does not matter, so it is combination. The number of line segments is $\binom{8}{2} = \boxed{28}$.

(b) No three vertices are collinear, so any combination of 3 vertices will make a triangle. Choosing 3 out of 8 is $\binom{8}{3} = \boxed{56}$.

(c) Let the length of a side of the cube be 1. First, let's consider a triangle that does not have an edge of the cube as any of its three sides. Then no two vertices of the triangle can be on the same edge of the cube. An example is ACF in the diagram to the right, and all such triangles are congruent to ACF. The side lengths are $\sqrt{2}$, $\sqrt{2}$, $\sqrt{2}$.

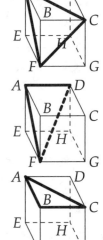

Next, if exactly one side of the triangle is an edge of the cube, then we can see that ADF is an example of such a triangle, and all such triangles are congruent to ADF. Its sides are 1, $\sqrt{2}$, $\sqrt{3}$.

Finally, a triangle cannot have all three of its sides be edges of the cube, so the only case left is where exactly two sides of the triangle are edges of the cube. An example is ABC, and all such triangles are congruent to ABC. Its sides are $1, 1, \sqrt{2}$.

Thus there are $\boxed{3}$ noncongruent triangles.

CHAPTER 5

More with Combinations

Exercises for Section 5.2

5.2.1 There are 5 steps to the right, and 2 steps up. These 7 steps can be made in any order, so the answer is $\binom{7}{2} = \frac{7 \times 6}{2 \times 1} = \boxed{21}$.

5.2.2 There are 4 steps to the right, and 6 steps down. These 10 steps can be made in any order, so the answer is $\binom{10}{4} = \frac{10 \times 9 \times 8 \times 7}{4 \times 3 \times 2 \times 1} = \boxed{210}$.

5.2.3

(a) There are 5 steps to the right, and 4 steps down. These 9 steps can be made in any order, so the answer is $\binom{9}{4} = \frac{9 \times 8 \times 7 \times 6}{4 \times 3 \times 2 \times 1} = \boxed{126}$.

(b) From E to F, it is 3 steps to the right and 1 step down, for a total of $\binom{4}{1} = \frac{4}{1} = 4$ different paths.

From F to G, it is 2 steps to the right and 3 steps down, for a total of $\binom{5}{2} = \frac{5 \times 4}{2 \times 1} = 10$ different paths. So there are $4 \times 10 = \boxed{40}$ paths from E to G that pass through F.

Exercises for Section 5.3

5.3.1 There are 8 Republicans and 3 spots for them, so there are $\binom{8}{3} = 56$ ways to choose the Republicans. There are 6 Democrats and 2 spots for them, so there are $\binom{6}{2} = 15$ ways to choose the Democrats. So there are $56 \times 15 = \boxed{840}$ ways to choose the subcommittee.

5.3.2

(a) We are choosing 6 starters from 14 players, which can be done in $\binom{14}{6} = \boxed{3003}$ ways.

(b) If all triplets are in the starting lineup, we are choosing the 3 remaining starters from 11 players, which can be done in $\binom{11}{3} = \boxed{165}$ ways.

(c) If exactly one of the triplets is in the lineup, we have 3 choices for which triplet to put in the starting lineup, and then 11 people to choose from for the remaining 5 spots. So the answer is $3 \times \binom{11}{5} = 3 \times 462 = \boxed{1386}$.

(d) We can add together the number of lineups with one triplet and with no triplets. The number of lineups with no triplets is $\binom{11}{6} = 462$, since we must choose 6 starters from the 11 remaining players. When one triplet is in the lineup, there are 1386 options, as found in part (c). So the total number of lineups with at most one triplet is $1386 + 462 = \boxed{1848}$.

5.3.3 Place Biter in the 3-dog group and Nipper in the 5-dog group. This leaves 8 dogs remaining to put in the last two spots of Biter's group, which can be done in $\binom{8}{2}$ ways. Then there are 6 dogs remaining for the last 4 spots in Nipper's group, which can be done in $\binom{6}{4}$ ways. The remaining 2-dog group takes the last 2 dogs. So the total number of possibilities is $\binom{8}{2} \times \binom{6}{4} = \boxed{420}$.

5.3.4 For every 3 different digits, there is one corresponding descending number, which is just the digits in descending order. So the answer is the number of combinations of three different digits, which is $\binom{10}{3} = \boxed{120}$.

5.3.5 We will break this into three cases.

Case 1: numbers of the form xyx ($x \neq 0$). Any pair of nonzero digits has a corresponding palindrome (xyx) mountain number, so the number of these is $\binom{9}{2} = 36$.

Case 2: numbers of the form xyz ($z \neq 0, x \neq z$). Any group of three nonzero digits ($y > x > z > 0$) has two corresponding mountain numbers (xyz and zyx), so the number of these is $2 \times \binom{9}{3} = 168$.

Case 3: numbers of the form xy0 ($x \neq 0, y \neq 0$). Any pair of nonzero digits has a corresponding mountain number in the form $xy0$, so there are $\binom{9}{2} = 36$ of these.

So the total number of mountain numbers is $36 + 168 + 36 = \boxed{240}$.

Exercises for Section 5.4

5.4.1

(a) There are 2 different boxes, so each of the 5 balls can be placed in two different locations. So the answer is $2^5 = \boxed{32}$.

(b) Since the boxes are indistinguishable, there are 3 possibilities for arrangements of the number of balls in each box.

Case 1: 5 balls in one box, 0 in the other box. We must choose 5 balls to go in one box, which can be done in $\binom{5}{5} = 1$ ways.

Case 2: 4 balls in one box, 1 in the other box. We must choose 4 balls to go in one box, which can be done in $\binom{5}{4} = 5$ ways.

Case 3: 3 balls in one box, 2 in the other box. We must choose 3 balls to go in one box, which can be done in $\binom{5}{3} = 10$ ways.

This gives us a total of $1 + 5 + 10 = \boxed{16}$ arrangements.

Also note that every arrangement of balls when the boxes are indistinguishable is counted twice in the distinguishable case. So we can simply divide the answer from part (a) by 2. However, this doesn't work if there's more than two boxes (as we'll see in the next problem).

(c) Since the balls are indistinguishable, we need only count the number of balls in the distinguishable boxes. We can put 5, 4, 3, 2, 1, or 0 ball in Box #1 (and the rest go in Box #2). So there are $\boxed{6}$ different arrangements.

(d) Since both the balls and boxes are indistinguishable, we can arrange them with 5 in one and 0 in the other, 4 in one and 1 in the other, or 3 in one and 2 in the other, for a total of $\boxed{3}$ different arrangements.

5.4.2

(a) There are 3 different boxes, so each of the 5 balls can be placed in three different locations. So the answer is $3^5 = \boxed{243}$.

(b) Since the boxes are indistinguishable, there are 5 different cases for arrangements of the number of balls in each box: $(5, 0, 0)$, $(4, 1, 0)$, $(3, 2, 0)$, $(3, 1, 1)$, or $(2, 2, 1)$.

$(5, 0, 0)$: There is only 1 way to put all 5 balls in one box.

$(4, 1, 0)$: There are $\binom{5}{4} = 5$ choices for the 4 balls in one of the boxes.

$(3, 2, 0)$: There are $\binom{5}{3} = 10$ choices for the 3 balls in one of the boxes.

$(3, 1, 1)$: There are $\binom{5}{3} = 10$ choices for the 3 balls in one of the boxes, and we simply split the last two among the other indistinguishable boxes.

$(2, 2, 1)$: There are $\binom{5}{2} = 10$ options for one of the boxes with two balls, then $\binom{3}{2} = 3$ options for the second box with two balls, and one option remaining for the third. However since the boxes with two balls are indistinguishable, we are counting each pair of balls twice, and must divide by two. So there are $\dfrac{10 \times 3}{2} = 15$ arrangements of balls as $(2, 2, 1)$.

Thus the total number of arrangements for 3 indistinguishable boxes and 5 distinguishable balls is $1 + 5 + 10 + 10 + 15 = \boxed{41}$. (Note that we cannot simply divide the answer from part (a) by $3! = 6$.)

(c) Since the balls are indistinguishable, we must only count the number of balls in the different boxes.

There are 3 ways to arrange the balls as $(5, 0, 0)$ (specifically, box 1 can have 5, box 2 can have 5, box 3 can have 5).

There are $3! = 6$ ways to arrange $(4, 1, 0)$ and $3! = 6$ ways to arrange $(3, 2, 0)$; in each case, we must choose one of the 3 boxes to have the largest number of balls, and also one of the remaining two boxes to be left empty.

However, there are only 3 ways to arrange $(3, 1, 1)$, and 3 ways to arrange $(2, 2, 1)$; in each case, we must choose one box to have the "different" number of balls (3 in the $(3, 1, 1)$ case and 1 in the $(2, 2, 1)$ case).

This gives a total of $3 + 6 + 6 + 3 + 3 = \boxed{21}$ arrangements.

(d) The ways to arrange indistinguishable balls into indistinguishable boxes only depends on the number of balls in the boxes. The ways to do this are $(5, 0, 0)$, $(4, 1, 0)$, $(3, 2, 0)$, $(3, 1, 1)$, $(2, 2, 1)$. There are $\boxed{5}$ ways.

(e) Clearly, we see that we cannot simply divide 3^5 by $3!$, since the result would not be an integer! The underlying reason that we cannot divide is that not all arrangements of distinguishable balls into distinguishable boxes permit $3!$ rearrangements of the boxes. For example, if we put all 5 balls into one box, then there are only 3 choices for which box contains the balls, not $3!$ choices, so when we make the boxes indistinguishable, we would have to divide these cases by 3, not by $3!$.

5.4.3

(a) There are n options for each of the 2 balls, so the answer is $n \times n = \boxed{n^2}$.

(b) The balls can be either placed together or separately in boxes since the boxes are indistinguishable, so there are $\boxed{2}$ ways.

(c) There are two cases: either the balls are in the same box or in separate boxes. If they are in the same box, there are n choices for which box it is. If they are in separate boxes, then this is the same as choosing 2 boxes out of n without regard to order, since the balls are not distinguishable; there are $\binom{n}{2}$ ways to do this. This gives a total of $n + \binom{n}{2} = n + \frac{n(n-1)}{2} = \boxed{\frac{n(n+1)}{2}}$ possibilities.

(d) Again, the balls can be either placed together or separately since both the balls and boxes are indistinguishable. There are $\boxed{2}$ ways.

Review Problems

5.10

(a) There are 5 steps to the right, and 4 steps up. These 9 steps can be made in any order, so the answer is $\binom{9}{4} = \boxed{126}$.

(b) There is 1 step to the right, and 2 steps up. These 3 steps can be made in any order, so the answer is $\binom{3}{1} = \boxed{3}$.

(c) There are 4 steps to the right, and 2 steps up. These 6 steps can be made in any order, so the answer is $\binom{6}{4} = \boxed{15}$.

(d) There are 3 paths from A to C and 15 paths from C to B, so there are $3 \times 15 = \boxed{45}$ paths from A to B that go through C.

5.11 We have 5 options for the choice of the vowel, and we must make 2 choices out of the remaining 21 letters, for a total of $\binom{21}{2} = 210$ choices for the consonants. This gives a total of $5 \times 210 = \boxed{1050}$.

5.12

(a) The socks must be either both white, both brown, or both blue. If the socks are white, there are $\binom{4}{2} = 6$ choices. If the socks are brown, there are $\binom{4}{2} = 6$ choices. If the socks are blue, there is $\binom{2}{2} = 1$ choice. So the total number of choices for socks is $6 + 6 + 1 = \boxed{13}$.

(b) If the socks are different, either white and brown, brown and blue, or white and blue can be picked. If the socks are white and brown, there are 4 options for the white sock and 4 options for the brown sock for a total of 16 choices. If the socks are brown and blue, there are 4 options for the brown sock and 2 options for the blue sock for a total of 8 choices. If the socks are white and blue, there are 4 options for the white sock and 2 options for the brown sock for a total of 8 choices. This gives a total of $16 + 8 + 8 = \boxed{32}$ choices.

Alternatively, since we can either draw two socks of the same color or two socks of different colors, we know that the number of matching pairs (13 from part (a)) subtracted from the total numbers of pairs of socks ($\binom{10}{2} = 45$) will give us the number of non-matching pairs. So $45 - 13 = 32$.

5.13

(a) With no restrictions, we are merely picking 6 students out of 14. This is $\binom{14}{6} = \boxed{3003}$.

(b) We are picking 3 boys out of 6, so there are $\binom{6}{3} = 20$ options for the boys on the team. We are picking 3 girls out of 8, so there are $\binom{8}{3} = 56$ options for the girls on the team. This gives a total of $20 \times 56 = \boxed{1120}$ choices.

(c) We do this problem similarly to part (b), except with three cases.

Case 1: 4 girls, 2 boys on the team. With 4 girls on the team, there are $\binom{8}{4} = 70$ ways to pick the girls, and $\binom{6}{2} = 15$ ways to pick the boys, for a total of $70 \times 15 = 1050$.

Case 2: 5 girls, 1 boy on the team. With 5 girls on the team, there are $\binom{8}{5} = 56$ ways to pick the girls, and $\binom{6}{1} = 6$ ways to pick the boy, for a total of $56 \times 6 = 336$.

Case 3: 6 girls on the team. With 6 girls on the team, there are $\binom{8}{6} = 28$ ways to pick the girls on the team.

This gives us a sum of $1050 + 336 + 28 = \boxed{1414}$.

5.14 To make the first committee, we choose 2 boys in $\binom{6}{2} = 15$ ways and 2 girls in $\binom{6}{2} = 15$ ways, for a total of $15 \times 15 = 225$ possible committees. Then to make the second committee, we choose 2 more boys in $\binom{4}{2} = 6$ ways and 2 more girls in $\binom{4}{2} = 6$ ways, for a total of $6 \times 6 = 36$ possible committees. So the number of ways to form two committees like this is $225 \times 36 = 8100$. But this counts each pair of committees twice (in other words, there is no "first" committee or "second" committee), so we must divide by 2 to get the final answer of $8100/2 = \boxed{4050}$ pairs of committees.

5.15

(a) There are $\binom{10}{5} = 252$ ways to choose players for the first team, and the second team gets the remaining players. However, since the teams are interchangeable, we must divide by two, so the answer is $252/2 = \boxed{126}$.

(b) If Steve and Danny are on the same team, there are 8 players to choose from for the other 3 spots on their team, so there are $\binom{8}{3} = \boxed{56}$ choices.

(c) If Steve and Danny are on opposite teams, there are 8 other players to choose from for the other 4 spots on Steve's team, so there are $\binom{8}{4} = \boxed{70}$ choices.

(d) The sum of the answers from parts (b) and (c) is the answer for part (a). This is because any intrasquad game either has Steve and Danny on the same team or on opposite teams.

5.16

(a) There are 3 options (boxes) for each of the 4 balls, so the number of ways is $3^4 = \boxed{81}$.

(b) Without regard to the distinguishability of the balls, they can be organized into groups of the following: (4,0,0),(3,1,0),(2,2,0),(2,1,1). Now we consider the distinguishability of the balls in each of these options.

(4,0,0): There is only 1 way to do this (since the boxes are indistinguishable).

(3,1,0): There are 4 options: we must pick the ball which goes into a box by itself.

(2,2,0): There are $\binom{4}{2} = 6$ ways to choose the balls for the first box, and the remaining go in the second box. However, the two pairs of balls are interchangeable, so we must divide by 2 to get $6/2 = 3$ arrangements.

(2,1,1): There are $\binom{4}{2} = 6$ options for picking the two balls to go in one box, and each of the other two balls goes into its own box.

The total number of arrangements is $1 + 4 + 3 + 6 = \boxed{14}$.

(c) Since the balls are indistinguishable, we only have to consider the number of balls in the boxes. The arrangements for balls in boxes are (4,0,0),(3,1,0),(2,2,0),(2,1,1). However, since the boxes are distinguishable, we must also consider the arrangement of balls in the boxes in order.

For (4,0,0), there are 3 different ways (box #1 can have 4, box #2 can have 4, or box #3 can have 4).

For (3,1,0), there are $3! = 6$ ways: we have 3 choices for the box containing 3 balls, then 2 choices for the box containing 1 ball.

For (2,2,0) there are 3 ways: we must choose the box which remains empty.

For (2,1,1) there are 3 ways: we must choose the box which gets 2 balls.

This gives a total of $3 + 6 + 3 + 3 = \boxed{15}$ arrangements.

(d) Since the balls and boxes are indistinguishable, we only need to consider the number of the balls in boxes without considering order. The arrangements are (4,0,0),(3,1,0),(2,2,0),(2,1,1), for a total of $\boxed{4}$ ways.

(e) We will consider this as a composite of two problems with two indistinguishable balls and 3 distinguishable boxes. For two indistinguishable green balls, we can place the balls in a box together or in separate boxes. There are 3 options to arrange them together (in box 1, 2, or 3) and 3 options for placing them separately (nothing in box 1, 2, or 3). So there are 6 ways to arrange the indistinguishable green balls. By the same reasoning, there are 6 ways to arrange the indistinguishable red balls, for a total of $6 \times 6 = \boxed{36}$ arrangements of the 4 balls.

Challenge Problems

5.17 If the edge from F to G were filled in, there would be $\binom{8}{5} = 56$ paths from D to E. However, we must subtract the paths that go through both F and G. There are $\binom{4}{2} = 6$ paths from D to F, 1 path from F to G, and $\binom{3}{2} = 3$ paths from G to E, which means there are $6 \times 1 \times 3 = 18$ paths from D to E that pass through F and G. So there are $56 - 18 = \boxed{38}$ from D to E that don't pass between F and G.

5.18 To maximize the number of intersections, we will draw the lines so that each line intersects with every other line (none are parallel), and so no three of them intersect at the same point. There are $\binom{9}{2} = 36$ pairs of lines, so there are $\boxed{36}$ points of intersections.

5.19

(a) The only way to split up the women is to have two women put in one group and one in the other. There are $\binom{3}{1} = \boxed{3}$ ways to pick the one woman to put in a group by herself. (Remember, these groups are indistinguishable.)

(b) The men must have 4 in one group and 3 in the other, since there must be at least one woman in each group. There are $\binom{7}{4} = 35$ ways to put the men in a group of 4, and the rest go in the other group of three, for a total of $\boxed{35}$ ways.

(c) There are 3 ways to split up the women and 35 ways to split up the men. The group of 3 men joins with the group of 2 women, and the group of 4 men joins with the third woman. So there are a total of $3 \times 35 = \boxed{105}$ to split up the teams.

5.20

(a) There are 5 desks to choose from for the first person, 4 choices for the second, and 3 choices for the third, so there are $5 \times 4 \times 3 = \boxed{60}$ ways to seat the students.

(b) There are 5 desks, and 3 to choose from that get a textbook, so there are $\binom{5}{3} = \boxed{10}$ choices of desks.

(c) We can put all the books on separate desks, and there are 10 arrangements for this (part b). We can put two of the books one desk (5 choices), and one of the books on another desk (4 choices), for a total of $5 \times 4 = 20$ arrangements. We can put all three books on one desk, and there are 5 desks to choose from for this. The sum of all these is $10 + 20 + 5 = \boxed{35}$ arrangements.

5.21 We do this by complementary counting; we will count the numbers with strictly increasing digits. We do this by cases, based on the number of digits in the number.

Case 1: one-digit numbers. All 9 one-digit numbers are increasing. So there are 9 of these.

Case 2: two-digit numbers. For every combination of two unique nonzero digits, there is exactly one increasing two-digit number. So the number of these is $\binom{9}{2} = 36$.

Case 3: three-digit numbers. For every combination of three unique nonzero digits (no number starts with zero), there is exactly one increasing three-digit number. So the number of these is $\binom{9}{3} = 84$.

So the total number of these is $9 + 36 + 84 = 129$. Since there are 999 numbers less than 1000 and 129 of them have their digits in strictly increasing order, there are $999 - 129 = \boxed{870}$ numbers with digits not strictly increasing.

5.22 20 men shake 38 hands each (not themselves or their spouses), so combined they shake 760 hands. 10 women shake only 19 hands (all the men but their husband), and 10 women shake hands with 28 people (all the men but their husbands and the 9 other friendly women), so women shake a total of $10 \times 19 + 10 \times 28 = 470$ hands. This means a total of 1230 hands have been shaken. However, each handshake has been counted twice, so we divide by two and see that there are $\boxed{615}$ handshakes.

5.23 Each circle can meet each other circle in at most two points. Draw the five circles such that each circle intersects with every other circle twice, and so that no three circles intersect at the same point.

Since there are $\binom{5}{2} = 10$ pairs of circles, and 2 intersections per pair of circles, there are $10 \times 2 = \boxed{20}$ intersections.

5.24 There must be at least as many groups of three friends as there are days of the year. Since the number of days is 365 and there are $\binom{n}{3}$ ways to choose a group of three friends out of n, we must find the smallest such n such that $\binom{n}{3} \geq 365$. Since $\binom{14}{3} = 364 < 365 < 455 = \binom{15}{3}$, there must be at least $\boxed{15}$ friends.

5.25 In this case $k = 4$, so a 4-digit number $a_1a_2a_3a_4$ is snakelike if $a_1 < a_2$, $a_2 > a_3$, and $a_3 < a_4$. So a_2 must be greater than both a_1 and a_3, and a_4 must be greater than a_3. Given any four distinct digits, we can see that a_2 must be either the highest or the second-highest of the four. If it is the second-highest, then a_1 and a_3 must be the two lowest digits, and a_4 is then the highest. If a_2 is the highest, then a_4 is either the second-highest (in which case a_1 and a_3 can be either of the lowest two digits), or a_4 is the third-highest (in which case a_3 must be the lowest).

So for any 4 distinct non-zero digits, there are 5 ways to arrange them to form a snakelike number. For example, only the following arrangements of the digits 1,2,3,4 are snakelike:

$$1324, \ 2314, \ 1423, \ 2413, \ 3412.$$

Since there are $\binom{9}{4}$ ways to choose 4 distinct nonzero digits, there are $5 \times \binom{9}{4} = 630$ snakelike 4-digit numbers with all nonzero digits.

If one of the digits is 0, then there are only 3 ways to arrange them to form a snakelike 4-digit number, since the first digit cannot be zero. There are $\binom{9}{3}$ ways to choose the other three nonzero digits, so there are $3 \times \binom{9}{3} = 252$ snakelike numbers with a digit of 0.

Therefore the number of snakelike 4-digit numbers is $630 + 252 = \boxed{882}$.

5.26 Mack must make 3 moves in the x-direction, 5 moves in the y-direction, and 2 moves in the z-direction. If all of these moves were distinguishable, then there would be 10! ways to arrange them. However, different moves in the same direction are not distinguishable, so we must divide by 3!, 5!, and 2!. So the number of ways Mack can arrive at $(3, 5, 2)$ is $\dfrac{10!}{3! \times 5! \times 2!} = \boxed{2,520}$.

CHAPTER 6

_____**Some Harder Counting Problems**

Challenge Problems

6.8

(a) (i) Since x, y, z are nonnegative integers that add up to 0, they must all be 0. There is only $\boxed{1}$ solution, $(0, 0, 0)$.

(ii) There are $\boxed{3}$ solutions, $(1, 0, 0), (0, 1, 0), (0, 0, 1)$.

(iii) We can look at cases based on the largest value among x, y, z. If the largest among x, y, z is 2, this forces the other two variables to be 0. There are 3 solutions, $(2, 0, 0), (0, 2, 0), (0, 0, 2)$. If the biggest value is 1, then exactly two are 1's and one is 0. There are 3 such solutions, $(1, 1, 0), (1, 0, 1), (0, 1, 1)$. There are a total of $\boxed{6}$ solutions.

(iv) We do cases again. If the largest is 3, the other two variables are 0's, and there are 3 such triples. If the largest is 2, the other two must be 0 and 1. There are $3! = 6$ such triples. If the largest is 1, all three variables must be 1, so there is 1 triple. There are $3 + 6 + 1 = \boxed{10}$ solutions.

(v) If the largest is 4, the other two must be 0, and there are 3 triples. If the largest is 3, the others are 0 and 1, and there are $3! = 6$ such triples. If the largest is 2, the other two can be either 2 and 0, or 1 and 1. Each case gives 3 triples. The largest cannot be less than 2. The final answer is $3 + 6 + 3 + 3 = \boxed{15}$ solutions.

(b) The pattern so far is $1, 3, 6, 10, 15$. Note that:

$$1 = 1$$
$$1 + 2 = 3$$
$$1 + 2 + 3 = 6$$
$$1 + 2 + 3 + 4 = 10$$
$$1 + 2 + 3 + 4 + 5 = 15$$

So we would expect that in the general case, the number of solutions to $x + y + z = n$ would be

$$1 + 2 + 3 + \cdots + (n + 1) = \frac{(n + 1)(n + 2)}{2} = \binom{n + 2}{2}.$$

(c) Once we know what $x + y$ is, then z must be $n - x - y$. So we need only to find the number of nonnegative x and y such that $x + y \leq n$. If $x + y = k$, there are $k + 1$ solutions, because x can be anything from 0 to k and y is fixed at $k - x$. So to get the number of solutions, we must add $k + 1$ for all k from 0 to n, which is equal to

$$1 + 2 + \cdots + (n + 1) = \frac{(n + 1)(n + 2)}{2} = \boxed{\binom{n + 2}{2}}.$$

6.9 Fix a vertex A of the 10-gon $ABCDEFGHIJ$ (with the vertices in that order). There are 4 diagonals that have their endpoints equidistant from vertex A (and equidistant from the vertex F opposite A), namely BJ, CI, DH and EG. These 4 are parallel to each other. There are 5 such sets of 4 parallel diagonals, since the vertices of the 10-gon can be paired into 5 sets. In each set, there are $\binom{4}{2} = 6$ pairs of parallel diagonals, so this gives a total of 30 pairs of parallel diagonals.

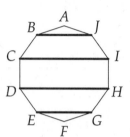

Now fix a side AB of the 10-gon. There are 3 diagonals that are parallel to this side AB (and to the side FG opposite AB), namely CJ, DI, and EH. We again have 5 such sets of 3 diagonals. In each set, there are $\binom{3}{2} = 3$ pairs of parallel diagonals, so the total for this case is 15 pairs.

The final answer is $30 + 15 = \boxed{45}$.

6.10 Since 30% of the objects in the bag are beads, 70% of the objects in the bag are coins. Half of the coins in the bag are gold, so half of the coins must be silver. Thus $\boxed{35\%}$ of the objects are silver coins.

6.11 The square has dimensions $n \times n$. Each diagonal has n tiles. If n is even, then the total number of blue tiles is $2n$, since the two diagonals don't intersect. If n is odd, then the total number of blue tiles is $n + n - 1 = 2n - 1$; we must subtract 1 because the center tile is counted twice, once for each diagonal. But we know that there are 121 blue tiles, which is an odd number, so $121 = 2n - 1$. This gives that the dimension of the square is $n = 61$, meaning there are $n \times n = 3721$ total tiles. Since all the other tiles are red, there are $3721 - 121 = \boxed{3600}$ red tiles.

6.12 We have a 4×4 grid of points, and we need to count how many nondegenerate triangles can we form. First, in how many ways can we pick three points without regard to whether they are collinear? That's a combination, given by $\binom{16}{3} = 560$. Out of these 560 triangles, how many are degenerate (equivalently, zero area or collinear vertices)? On each horizontal or vertical line, there are $\binom{4}{3} = 4$ degenerate triangles, and there are 8 such lines (4 horizontal and 4 vertical), for a total of 32 such degenerate triangles. The only other ways to have collinear points is on lines of slope 1 or -1. We have three lines of slope 1 which contain at least 3 of our grid points, two containing 3 points and one containing 4 points, thus there are $2\binom{3}{3} + \binom{4}{3} = 6$ sets of 3 collinear points in our grid lying on a line of slope 1. The same is true for the lines of slope -1. So the total number of triangles with positive area is $560 - 32 - 6 - 6 = \boxed{516}$.

6.13 We divide into cases based on the genders of the people at the table.

Case 1: 5 women, 0 men. Everybody is sitting next to a woman and of course no one is sitting next to a man, so the only possibility for the ordered pair is $(5, 0)$.

Case 2: 4 women, 1 man. Everyone sits next to some woman, and exactly two sit next to the man, so the ordered pair must be $(5,2)$.

Case 3: 3 women, 2 men. There are two subcases: either two men sit next to each other or they don't. The former case gives $(5,4)$, while the latter gives $(4,3)$.

Case 4: 2 women, 3 men. This is analogous to the case where there are 3 women and 2 men, except the men and women are reversed. The possible ordered pairs are $(4,5)$ and $(3,4)$.

Case 5: 1 woman, 4 men. This is the reverse of Case 2, so the ordered pair is $(2,5)$.

Case 6: 0 women, 5 men. This is the reverse of Case 1, so the ordered pair is $(0,5)$.

The number of possible ordered pairs is $\boxed{8}$.

6.14 When you look from a point so that you can see 3 faces, the number of unit cubes visible is maximized. In that situation, the $9 \times 9 \times 9$ cube that is behind the 3 visible faces is hidden from the view. The number of visible cubes is thus $10^3 - 9^3 = \boxed{271}$.

6.15

(a) First we place the 4 X's in the last 8 slots, which we can do in $\binom{8}{4}$ ways. We can then place the 4 Y's and the 4 Z's in the 8 remaining slots in $\binom{8}{4}$ ways. So the number of arrangements is $\binom{8}{4} \times \binom{8}{4} = 70 \times 70 = \boxed{4900}$.

(b) We do casework, based on how many of the X's end up in the middle 4 letters.

Case 1: All 4 X's are in the middle 4 slots. There is only 1 way to do this, and then there are $\binom{8}{4} = 70$ ways to place the Y's, for a total of $1 \times 70 = 70$ arrangements.

Case 2: 3 X's in the middle 4 slots, 1 X in the last 4 slots. There are $\binom{4}{3} \times 4 = 16$ ways to do this, and then there are $\binom{7}{4} = 35$ ways to place the Y's, for a total of $16 \times 35 = 560$ arrangements.

Case 3: 2 X's in the middle 4 slots, 2 X's in the last 4 slots. There are $\binom{4}{2} \times \binom{4}{2} = 36$ ways to do this, and then there are $\binom{6}{4} = 15$ ways to place the Y's, for a total of $36 \times 15 = 540$ arrangements.

Case 4: 1 X in the middle 4 slots, 3 X's in the last 4 slots. There are $4 \times \binom{4}{3} = 16$ ways to do this, and then there are $\binom{5}{4} = 5$ ways to place the Y's, for a total of $16 \times 5 = 80$ arrangements.

Case 5: All 4 X's are in the last 4 slots. There is only 1 way to do this, and then there is only one way to place the Y's (they must go in the first 4 slots), so there is only 1 arrangement.

Therefore there are $70 + 560 + 540 + 80 + 1 = \boxed{1251}$ total arrangements.

(c) As in part (b), we do casework, based on how many of the X's end up in the middle 4 letters.

Case 1: All 4 X's are in the middle 4 slots. There is only 1 way to do this, and then we must place the four Z's in the first four slots and the four Y's in the last four slots. So there is only 1 arrangement.

Case 2: 3 X's in the middle 4 slots, 1 X in the last 4 slots. There are $\binom{4}{3} \times 4 = 16$ ways to do this. We must then place a Z in the remaining middle slot and 3 Y's in the remaining final slots. We then have 3 Z's and 1 Y left for the first four slots, which can be arranged in 4 ways. So there are $16 \times 4 = 64$ arrangements.

Case 3: 2 X's in the middle 4 slots, 2 X's in the last 4 slots. There are $\binom{4}{2} \times \binom{4}{2} = 36$ ways to do this. We must then place 2 Z's in the remaining middle slots and 2 Y's in the remaining final slots. We then have 2 Z's and 2 Y's left for the first four slots, which can be arranged in $\binom{4}{2} = 6$ ways. So there are $36 \times 6 = 216$ arrangements.

Case 4: 1 X in the middle 4 slots, 3 X's in the last 4 slots. There are $4 \times \binom{4}{3} = 16$ ways to do this. We must then place 3 Z's in the remaining middle slots and 1 Y in the remaining final slot. We then have 1 Z and 3 Y's left for the first four slots, which can be arranged in 4 ways. So there are $16 \times 4 = 64$ arrangements.

Case 5: All 4 X's are in the last 4 slots. There is only 1 way to do this, and then we must place the four Y's in the first four slots and the four Z's in the four middle slots, so there is only 1 arrangement.

Therefore there are $1 + 64 + 216 + 64 + 1 = \boxed{346}$ total arrangements.

6.16 Note that in x, the digit 1 occurs ten times in the units place (in 1, 11, 21, ..., 91) and ten times in the tens place (in 10, 11, 12, ..., 19). The same is true for 3, 5, 7, 9 also. Therefore $x = 20(1 + 3 + 5 + 7 + 9) + O(100) = 500 + 1 = \boxed{501}$. As for y, each of the digits 2, 4, 6, 8 occurs 20 times, just as for the odd digits. Thus $y = 20(2 + 4 + 6 + 8) = \boxed{400}$.

6.17 To get to $(3, 5, 2)$, each path corresponds to a 10-letter string where each letter is X, Y, or Z (where X indicates a step in the x-direction, Y indicates a step in the y-direction, and Z indicates a step in the z-direction). The number of such strings is $\dfrac{10!}{3!5!2!} = 2520$. The number of paths that stop at $(1, 3, 1)$ is the number of ways to get from $(0,0,0)$ to $(1,3,1)$ multiplied by the number of ways to get from $(1,3,1)$ to $(3,5,2)$, which is $\dfrac{5!}{1!3!1!} \times \dfrac{5!}{2!2!1!} = 20 \times 30 = 600$. We want to exclude these paths, so the answer is the difference of the two, which is $2520 - 600 = \boxed{1920}$.

6.18 Notice how once he picks a column, he has no choice on which clay target within the column he can shoot, since he must shoot the bottom one. So his shooting sequence corresponds exactly with his choice of which columns to shoot at in each turn. If we let L denote shooting at the left column, C denote shooting at the center column, and R denote shooting at the right column, then the number of ways that the marksman can shoot the targets in equal to the number of ways to arrange 3 R's, 3 C's, and 2 L's in a word. The number of such arrangements is $\dfrac{8!}{3!3!2!} = \boxed{560}$.

6.19 We can break the count up into cases.

Case 1: rectangles with height 1. This must either be an entire row, or a row omitting the white square at the end. So there are 2 rectangles for each row, for a total of 16.

Case 2: rectangles with width 1. This is the same as Case 1, but with columns instead of rows. So there are 16 here as well.

Case 3: rectangles with both length and width at least 2. It's fairly easy to count these: we choose 2 (out of 8) rows and 2 (out of 8) columns to serve as the boundaries, so there are $\binom{8}{2}\binom{8}{2} = 28^2 = 784$ rectangles. But now we need to exclude those which don't contain at least 4 black squares. All of them contain at least 2 black squares, so we need to exclude those that contain exactly 2 or exactly 3.

The only rectangles in Case 3 which contain exactly 2 black squares are the 2×2 rectangles, and there are $7 \times 7 = 49$ of these.

The only rectangles in Case 3 which contain exactly 3 black squares are the 2×3 and 3×2 rectangles, and there are $(6 \times 7) + (7 \times 6) = 84$ of these.

So there are $784 - 49 - 84 = 651$ rectangles in Case 3 which contain at least 4 black squares.

Therefore, there are a total of $16 + 16 + 651 = \boxed{683}$ desired rectangles.

6.20   There are 2 paths to each T which is diagonally adjacent to the center M, and only one path to each T which is horizontal or vertical to the center M. From the "diagonal" T's, there are 2 choices of H's to complete the path, and from the other T's, there are 3 choices of H's. So the total number of paths is $(4 \times 2 \times 2) + (4 \times 1 \times 3) = 16 + 12 = \boxed{28}$.

CHAPTER 7

Introduction to Probability

Exercises for Section 7.2

7.2.1

(a)   Rolling a ⚅ is 1 out of 6 possible outcomes, so its probability is $\boxed{\dfrac{1}{6}}$.

(b)   A ⚀, ⚂, or ⚄ can be rolled, which is 3 out of 6 possible outcomes, so its probability is $\dfrac{3}{6} = \boxed{\dfrac{1}{2}}$.

(c)   A ⚀ or ⚁ can be rolled for success, which is 2 out of 6 possible outcomes, so its probability is $\dfrac{2}{6} = \boxed{\dfrac{1}{3}}$.

7.2.2

(a)   There are 13 ♡'s and 52 cards total, so the probability that the top card is a ♡ is $\dfrac{13}{52} = \boxed{\dfrac{1}{4}}$.

(b)   There are four 5's and 52 cards total, so the probability that the top card is a 5 is $\dfrac{4}{52} = \boxed{\dfrac{1}{13}}$.

(c)   There is one King of ◇ and 52 cards total, so the probability that the top card is a King of ◇ is $\boxed{\dfrac{1}{52}}$.

(d)   There are $3 \times 4 = 12$ face cards and 52 cards total, so the probability that the top card is a face card is $\dfrac{12}{52} = \boxed{\dfrac{3}{13}}$.

(e)   There are 26 ways to choose the first card to be red, then 26 ways to choose the second card to be black. There are 52×51 ways to choose any two cards. So the probability is $\dfrac{26 \times 26}{52 \times 51} = \boxed{\dfrac{13}{51}}$.

(f)   There are 4 ways to choose the first card to be a 3, then 4 ways to choose the second card to be an 8. There are 52×51 ways to choose any two cards. So the probability is $\dfrac{4 \times 4}{52 \times 51} = \boxed{\dfrac{4}{663}}$.

(g) There are 4 ways to choose the first card to be an Ace, then 3 ways to choose the second card to be another Ace. There are 52×51 ways to choose any two cards. So the probability is $\dfrac{4 \times 3}{52 \times 51} = \boxed{\dfrac{1}{221}}$.

(h) There are 13 ways to choose the first card to be a ♠, then 12 ways to choose the second card to be another ♠, then 11 ways to choose the third card to be a ♠. There are $52 \times 51 \times 50$ ways to choose any three cards. So the probability is $\dfrac{13 \times 12 \times 11}{52 \times 51 \times 50} = \boxed{\dfrac{11}{850}}$.

7.2.3 In all four parts, there are $2^4 = 16$ possible outcomes, since each of the 4 coins can land 2 different ways (heads or tails).

(a) There is only 1 way that they can all come up heads, so the probability of this is $\boxed{\dfrac{1}{16}}$.

(b) There are 2 possibilities for the dime and 2 for the quarter, so there are $2 \times 2 = 4$ successful outcomes, and the probability of this is $\dfrac{4}{16} = \boxed{\dfrac{1}{4}}$.

(c) There are 2 possibilities for the penny and the dime: either they're both heads or they're both tails. There are also 2 possibilities for the nickel and 2 possibilities for the quarter. So there are $2 \times 2 \times 2 = 8$ successful outcomes, and the probability of success is $\dfrac{8}{16} = \boxed{\dfrac{1}{2}}$.

(d) If the quarter is heads, there are 8 possibilities, since each of the other three coins may come up heads or tails. If the quarter is tails, then the nickel and dime must be heads, so there are 2 possibilities, since the penny can be heads or tails. So there are $8 + 2 = 10$ successful outcomes, and the probability of success is $\dfrac{10}{16} = \boxed{\dfrac{5}{8}}$.

Exercises for Section 7.3

7.3.1 There are 3 yellow faces and 8 faces total, so the probability of rolling a yellow face is $\boxed{\dfrac{3}{8}}$.

7.3.2 There are 3 ways to roll a 4, ⚁ on the first die and ⚀ on the second die, ⚂ on the first die and ⚀ on the second die, and ⚀ on the first die and ⚂ on the second die. There are 36 total possibilities, so the probability is $\dfrac{3}{36} = \boxed{\dfrac{1}{12}}$.

7.3.3 There are 13 integers between 5 and 17 inclusive, so there are $\binom{13}{2} = 78$ ways to choose two of them without regard to order. In order for the product of two integers to be odd, both of the integers themselves must be odd. There are 7 odd integers between 5 and 17 inclusive, so there are $\binom{7}{2} = 21$ ways to choose two of them without regard to order. Therefore, the desired probability is $\dfrac{21}{78} = \boxed{\dfrac{7}{26}}$.

7.3.4 In this set of integers, there are 5 tens digits: {2,3,4,5,6}. If 5 integers all have different tens digits, then there must be exactly one integer among the 5 with each tens digit. Since there are 10 different integers for each tens digit, the number of ways to pick, without regard to order, 5 different integers with different tens digits is 10^5. The total number of combinations of 5 integers is $\binom{50}{5}$. So the probability that 5 integers drawn all have the different tens digits is

$$\frac{10^5}{\binom{50}{5}} = \boxed{\frac{2500}{52969}}.$$

Exercises for Section 7.4

7.4.1 There are $\binom{10}{2} = 45$ ways to choose two members of the group, and there are $\binom{5}{2} = 10$ ways to choose two girls. Therefore, the probability that two members chosen at random are girls is $\frac{10}{45} = \boxed{\frac{2}{9}}$.

7.4.2 There are $\binom{20}{2} = 190$ ways to choose two members of the group. There are 12 ways to choose a boy and 8 ways to choose a girl for a total of $12 \cdot 8 = 96$ ways to choose a boy and a girl. This means that there is a $\frac{96}{190} = \boxed{\frac{48}{95}}$ chance that the two random members of the group are a boy and a girl.

7.4.3 In the word SEVEN, there are 5 letters to be arranged with 2 E's, so there are $\frac{5!}{2!} = 60$ arrangements of SEVEN. If the E's must be adjacent, we can treat this as an arrangement of the word SVN(EE), where the E's are one letter. There are $4! = 24$ ways to arrange this. Therefore, the probability that a random arrangement of SEVEN has both E's next to each other is $\frac{24}{60} = \boxed{\frac{2}{5}}$.

7.4.4 Solution 1: There are $\binom{8}{2} = 28$ ways to choose two vertices of an octagon. There are 8 ways to choose a pair of adjacent vertices (vertices 1 and 2, 2 and 3, ..., 8 and 1). So the probability that 2 vertices chosen at random are adjacent is $\frac{8}{28} = \boxed{\frac{2}{7}}$.
Solution 2: Choose the first vertex. There are 2 vertices adjacent and 5 that are not, so the chance that the second vertex is adjacent is $\boxed{\frac{2}{7}}$.

7.4.5 There are $\binom{52}{3} = 22{,}100$ ways to choose 3 cards out of 52, without regard to order. To choose two cards of matching rank, there are 13 different ranks and $\binom{4}{2} = 6$ combinations of suits to choose from, for a total of $13 \times 6 = 78$ different possibilities. There are 48 remaining cards not in the same rank as the first two. This means there are $78 \times 48 = 3{,}744$ ways to choose a hand that is a pair. So the probability that a randomly drawn hand is a pair is $\frac{3744}{22100} = \boxed{\frac{72}{425}}$.

7.4.6 There are $\binom{7}{2} = 21$ pairs of points in the heptagon, and all but 7 (the sides of the heptagon) are diagonals, which means there are 14 diagonals. So there are $\binom{14}{2} = 91$ pairs of diagonals. Any four points on the heptagon uniquely determine a pair of intersecting diagonals. (If vertices A, B, C, D are chosen, where $ABCD$ is a convex quadrilateral, the intersecting pair of diagonals are AC and BD.) So the number of sets of intersecting diagonals is the number of combinations of 4 points, which is $\binom{7}{4} = 35$.

So the probability that a randomly chosen pair of diagonals intersect is $\frac{35}{91} = \boxed{\frac{5}{13}}$.

Review Problems

7.11

(a) There are 5 white balls and 11 balls total, which means there is a $\boxed{\frac{5}{11}}$ probability that the ball drawn out will be white.

(b) There are $\binom{11}{2} = 55$ combinations of two balls that can be drawn. There are $\binom{5}{2} = 10$ combinations of two white balls that can be drawn. So the probability that two balls pulled out are both white is $\frac{10}{55} = \boxed{\frac{2}{11}}$.

(c) There are $\binom{11}{5} = 462$ ways to choose 5 balls out of the box. There is only $\binom{5}{5} = 1$ way to choose 5 white balls out of 5. This means that the probability that all 5 balls are white is $\boxed{\frac{1}{462}}$.

7.12 In all of these parts, there are $6 \times 6 = 36$ possible rolls of the dice.

(a) There are 6 different ways to roll doubles ($\boxed{\cdot\ \cdot}$, $\boxed{\cdot}\ \boxed{\cdot}$, ..., $\boxed{\vdots\ \vdots}$), which means the probability of rolling doubles is $\frac{6}{36} = \boxed{\frac{1}{6}}$.

(b) There are 4 different ways to roll a 9 ($\boxed{\cdot\cdot\ \vdots\vdots}$, $\boxed{\vdots\ \vdots\cdot\cdot}$, $\boxed{\cdot\cdot\vdots\ \vdots}$, $\boxed{\vdots\vdots\ \cdot\cdot}$), which makes the probability of rolling a 9 equal to $\frac{4}{36} = \boxed{\frac{1}{9}}$.

(c) We count by casework the number of acceptable outcomes.

Case 1: the first die is a $\boxed{\cdot}$. This means the second die can be either $\boxed{\cdot\cdot}$, $\boxed{\vdots}$, or $\boxed{\cdot\cdot}$. So there are 3 ways.

Case 2: the first die is a $\boxed{\cdot\ \cdot}$. This means the second die can be either $\boxed{\cdot}$, $\boxed{\cdot\cdot}$, or $\boxed{\vdots}$. So there are 3 ways.

Case 3: the first die is a $\boxed{\cdot\cdot}$. This means the second die can be either $\boxed{\cdot}$, $\boxed{\cdot\ \cdot}$, or $\boxed{\cdot\cdot}$. So there are 3 ways.

Case 4: the first die is a $\boxed{\vdots\vdots}$. This means the second die can be either $\boxed{\cdot}$ or $\boxed{\cdot\ \cdot}$. So there are 2 ways.

Case 5: the first die is a $\boxed{\cdot\cdot\cdot}$. This means the second die can only be $\boxed{\cdot}$. So there is 1 way.

Case 6: the first die is a $\boxed{\vdots\ \vdots}$. No value of the second die will give a total less than 7.

The number of ways to roll greater than 3 but less than 7 is $3 + 3 + 3 + 2 + 1 = 12$, which means the chance of rolling this is $\frac{12}{36} = \boxed{\frac{1}{3}}$.

(d) Solution 1: There are 6 ways that the first roll can be a ⚀: (⚀⚁, ⚀⚂, ..., ⚀⚅), and there are 6 ways that the second roll can be a ⚀: (⚁⚀, ⚂⚀, ..., ⚅⚀). However, we counted twice when they are both ⚀, so the number of ways that one of the two dice is ⚀ is $6 + 6 - 1 = 11$. This means that the probability that at least one of the dice is ⚀ is $\boxed{\dfrac{11}{36}}$.

Solution 2: There are 5 ways in which the first roll is not ⚀, and 5 ways in which the second roll is not ⚀, so there are $5 \times 5 = 25$ ways in which neither die shows ⚀. Therefore there are $36 - 25 = 11$ ways in which one or both dice show ⚀. So the probability of this is $\boxed{\dfrac{11}{36}}$.

7.13 For all of these parts, there are 100 numbers possible between 1 and 100.

(a) There are 10 perfect squares between 1 and 100: $1^2, 2^2, \ldots, 10^2$. So the probability that a randomly selected number is a perfect square is $\dfrac{10}{100} = \boxed{\dfrac{1}{10}}$.

(b) There are 33 multiples of 3 between 1 and 100: $(3, 6, 9, \ldots, 99) = (1 \times 3, 2 \times 3, 3 \times 3, \ldots, 33 \times 3)$. So the probability that a randomly selected number is a multiple of 3 is $\boxed{\dfrac{33}{100}}$.

(c) There are 6 divisors of 50: 1,2,5,10,25,50. So the probability that a randomly selected number is a divisor of 50 is $\dfrac{6}{100} = \boxed{\dfrac{3}{50}}$.

7.14 There are 2 possible outcomes for the coin and 6 possible outcomes for the die, so there are $2 \times 6 = 12$ equally likely outcomes. Only 1 of these is a successful outcome: the coin must show heads and the die must show ⚂. So the probability is $\boxed{\dfrac{1}{12}}$.

7.15 Let B denote drawing a black ball and W denote drawing a white ball. There are two possible successful orders: $BWBWBWBW$ or $WBWBWBWB$. There are $\binom{8}{4} = 70$ ways to arrange four B's and four W's, so the probability that a random arrangement is successful is $\dfrac{2}{70} = \boxed{\dfrac{1}{35}}$.

7.16

(a) There are $\binom{13}{3} = 286$ ways to choose a subcommittee from the committee, and $\binom{5}{3} = 10$ ways to choose a subcommittee of all Republicans. The chance that a random subcommittee is all Republican is $\dfrac{10}{286} = \boxed{\dfrac{5}{143}}$.

(b) There are 5 ways to choose a Republican, 6 ways to choose a Democrat, and 2 ways to choose an Independent for a total of $5 \times 6 \times 2 = 60$ different subcommittees of a Republican, Democrat, and Independent. So the probability that the subcommittee is made up of a Republican, Democrat, and Independent is $\dfrac{60}{286} = \boxed{\dfrac{30}{143}}$.

7.17 There are $6 \times 6 = 36$ possible outcomes. The only way that they can roll an odd product is if both their rolls are odd. Since 3 of the 6 faces on each die are odd, this can occur in $3 \times 3 = 9$ ways. So an even product can occur in $36 - 9 = 27$ ways, and the probability is thus $\dfrac{27}{36} = \boxed{\dfrac{3}{4}}$.

Challenge Problems

7.18   In any series of 7 flips, either at least 4 heads or at least 4 tails appear. For every possible series of flips with at least 4 heads, the exact opposite series of flips (for example, HHTHHTH and TTHTTHT) has at least 4 tails, and therefore doesn't have 4 heads. So we can pair every series of flips with at least 4 heads with exactly one series of flips with fewer than 4 heads, which means $\boxed{1/2}$ of the series of flips have at least 4 heads.

7.19

(a)   There are 30 multiples of 2 between 1 and 60: $(2, 4, 6, \ldots, 60) = (1 \times 2, 2 \times 2, \ldots, 30 \times 2)$. So the probability that a multiple of 2 is selected is $\dfrac{30}{60} = \boxed{\dfrac{1}{2}}$.

(b)   There are 20 multiples of 3 between 1 and 60: $(3, 6, 9, \ldots, 60) = (1 \times 3, 2 \times 3, \ldots, 20 \times 3)$. So the probability that a multiple of 3 is selected is $\dfrac{20}{60} = \boxed{\dfrac{1}{3}}$.

(c)   In every group of 6 consecutive numbers (e.g. $1, 2, 3, 4, 5, 6$), there are 4 multiples of 2 or 3 (e.g. $2, 3, 4, 6$). Since there are 10 sets of 6 consecutive numbers (1–6, 7–12, $\ldots$, 55–60), there are $10 \cdot 4 = 40$ multiples of 2 or 3 between 1 and 60. So the probability that a multiple of 2 or 3 is selected is $\dfrac{40}{60} = \boxed{\dfrac{2}{3}}$.

(d)   Some numbers are both multiples of 2 and multiples of 3, so if we were to add the answers from part (a) and part (b), we would be counting numbers that are multiples of 2 and 3 twice.

(e)   Any number that is a multiple of 2 and 3 is a multiple of 6. Since the numbers that are multiples of 6 are counted twice, we can subtract the number of multiples of 6 so that they are only counted once. Since there are 10 multiples of 6 between 1 and 60, the number of multiples of 2 or 3 between 1 and 60 is $30 + 20 - 10 = \boxed{40}$.

7.20   If Paco's first number is 1 or 2, then any of the 10 numbers can appear on the second spinner, so this gives 10 successful spins for either 1 or 2 appearing on the first spin. If Paco's first number is 3, then any number other than 10 can appear, so there are 9 successful spins. If his first number is 4, then the second spinner must show 1 through 7 (inclusive), so there are 7 more successful spins. Finally, if Paco's first number is 5, then the second spinner must show 1 through 5 (inclusive). So there are a total of $10 + 10 + 9 + 7 + 5 = 41$ successful spins, and $5 \times 10 = 50$ possible spins, giving a probability of $\boxed{41/50}$.

7.21   We will first count the number of hands which are three of a kind. There are 13 possible ranks that the three cards of the same rank could be, and there are $\binom{4}{3} = 4$ ways to choose the suits. There are $\binom{12}{2} = 66$ ways to choose ranks for the final two cards, and there are $4 \times 4 = 16$ ways to choose suits for those two cards. So the total number of ways to draw three of a kind is $13 \times 4 \times 66 \times 16 = 54{,}912$. Since there are $\binom{52}{5} = 2{,}598{,}960$ different 5-card hands that can be drawn, the probability of three of kind is

$$\frac{54{,}912}{2{,}598{,}960} = \boxed{\frac{88}{4165}} \approx .021.$$

Now we will count the number of hands which are two pairs. There are $\binom{13}{2} = 78$ ways to choose the two ranks for the pairs, and $\binom{4}{2} = 6$ ways to choose the suits for each rank, for a total of $6 \times 6 = 36$ ways to choose the suits. This leaves 44 cards of the 11 remaining ranks to choose from for the fifth card. This makes the number of different two pairs hands $78 \times 36 \times 44 = 123{,}552$, which means the probability of two pairs is $\frac{123{,}552}{2{,}598{,}960} = \boxed{\frac{198}{4165}} \approx .048$. So getting two pairs is more than twice as likely as getting three of a kind.

Intuitively, if you were to have (for example) two kings, a queen, and a jack as your first four cards, then it would be much more likely to draw a second queen or jack (since there are 6 queens and jacks remaining in the deck) than a third king (since there are only 2 remaining in the deck). This makes the probability of getting two pair much higher. Of course, this doesn't take into consideration the likelihood of making two pairs or three of a kind in your first four cards, but it does give a general idea.

7.22

(a) We can make a table of the possible outcomes:

	⚀	⚁	⚂	⚃	⚄	⚅
⚀	2	3	4	5	3	7
⚄	6	7	8	9	7	11
⚂	4	5	6	7	5	9
⚃	5	6	7	8	6	10
⚄	6	7	8	9	7	11
⚅	7	8	9	10	8	12

There are 8 sevens in the chart, so the probability of rolling a seven is $\frac{8}{36} = \boxed{\frac{2}{9}}$.

(b) There are sixteen possible even rolls, so the probability of rolling an even number is $\frac{16}{36} = \boxed{\frac{4}{9}}$.

7.23 Let the probability that a girl is chosen be x. Note that the probability that a girl is chosen is the same as the fraction of girls in the group. This means that the probability that a boy is chosen is $\frac{3}{4}x$. Since either a boy or girl must be chosen, the sum of these probabilities must be one. So $x + \frac{3}{4}x = 1$, which means $x = \boxed{\frac{4}{7}}$.

7.24 We choose the first block at random. There are $\binom{4}{2} = 6$ different ways that the second block can share two of the four properties with the first block, so there are 6 exclusive cases.

Case 1: Same material and color. The other block must have a different size and different shape. There are are 3 choices for the different size and 4 choices for the different shape for a total of $3 \times 4 = 12$ blocks with the same material and color only.

Case 2: Same material and size. The other block must have a different color and different shape. There are 2 choices for the different color and 4 choices for the different shape for a total of $2 \times 4 = 8$ with the same material and size only.

Case 3: Same material and shape. The other block must have a different color and different size. There are 2 choices for the different color and 3 choices for the different size for a total of $2 \times 3 = 6$ with the same material and shape only.

Case 4: Same color and size. The other block must have a different material and different shape. There is 1 choice for the different material and 4 choices for the different shape for a total of $1 \times 4 = 4$ with the same color and size only.

Case 5: Same color and shape. The other block must have a different material and different size. There is 1 choice for the different material and 3 choices for the different size for a total of $1 \times 3 = 3$ with the same color and shape only.

Case 6: Same size and shape. The other block must have a different material and different color. There is 1 choice for the different material and 2 choices for the different color for a total of $1 \times 2 = 2$ with the same size and shape only.

Summing all these cases, there are 35 blocks with exactly two characteristics in common with the original block. There are a total of 119 other blocks, so the probability that the two blocks have exactly two characteristics in common is $\dfrac{35}{119} = \boxed{\dfrac{5}{17}}$.

CHAPTER 8

Basic Probability Techniques

Exercises for Section 8.2

8.2.1 The first case is that we have 1 white marble. There are a total of 4 marbles, so we can choose two marbles in $\binom{4}{2} = 6$ ways. We can choose two marbles of the same color only by choosing two red marbles out of the 3 total red marbles, in $\binom{3}{2} = 3$ ways. So the probability in this case is $\dfrac{3}{6} = \boxed{\dfrac{1}{2}}$.

The second case is that we have 6 white marbles. There are a total of 9 marbles, so we can choose two marbles in $\binom{9}{2} = 36$ ways. We can choose two marbles of the same color by choosing two red marbles from the 3 total red marbles, in $\binom{3}{2} = 3$ ways; we can also choose two white marbles from the 6 total white marbles, in $\binom{6}{2} = 15$ ways. So the probability in this case is $\dfrac{3 + 15}{36} = \boxed{\dfrac{1}{2}}$.

8.2.2 Rolling two dice has $6 \times 6 = 36$ possible outcomes. The only perfect squares that we can roll are 4 and 9. Pairs adding up to 4 are ⚀⚂, ⚁⚁, and ⚂⚀. Those that add up to 9 are ⚂⚅, ⚃⚄, ⚄⚃, and ⚅⚂. The answer is $\boxed{\dfrac{7}{36}}$.

8.2.3 There are 16 odd-numbered cards, namely 4 suits for each of the 4 odd digits. There are 13 ♠s, but 4 of these we already counted among the odd-numbered cards. So the total number of cards that are odd or a ♠ is $16 + (13 - 4) = 25$, and the probability is $\boxed{\dfrac{25}{52}}$.

8.2.4 The number of possible outcomes after flipping 3 coins is $2 \times 2 \times 2 = 8$. If the dime isn't heads, we need both the penny and nickel to be heads. If the dime is heads, the other two could be anything. So there are a total of $1 + (2 \times 2) = 5$ ways for us to get at least 6¢. The answer is $\boxed{\dfrac{5}{8}}$.

8.2.5 The number of ways for the outcome to have exactly 0, 1, or 2 heads are $\binom{8}{0} = 1$, $\binom{8}{1} = 8$, or $\binom{8}{2} = 28$, respectively. There are 2^8 total possible outcomes (2 possibilities for each coin, and 8 coins). So the answer is $\dfrac{1 + 8 + 28}{2^8} = \boxed{\dfrac{37}{256}}$.

8.2.6 The number of ways to choose one white ball and one black ball is $5k$, since there are 5 choices for the white ball and k choices for the black ball. The number of ways to pick any 2 balls out of $(k+5)$ balls is $\binom{k+5}{2} = \frac{(k+5)(k+4)}{2}$. So we have to solve for k in the equation

$$\frac{5k}{\frac{(k+5)(k+4)}{2}} = \frac{10}{21}.$$

After clearing the denominators, we can simplify to $210k = 10(k+5)(k+4)$, giving the quadratic $10k^2 - 120k + 200 = 0$. This is the same as $k^2 - 12k + 20 = 0$, which factors as $(k-2)(k-10) = 0$, so its solutions are $k = \boxed{2}$ or $k = \boxed{10}$. Both are valid solutions to this problem.

8.2.7 Since each coin has 2 possible outcomes, there are 2^n possible outcomes for the n coins. The number of outcomes in which the number of tails is 0 or 1 is $\binom{n}{0} + \binom{n}{1} = 1 + n$. So the probability of having at most one tail is $\frac{1+n}{2^n}$. Therefore, we must solve the equation

$$\frac{1+n}{2^n} = \frac{3}{16}.$$

We can check (simply by plugging in values of n) that if $1 \le n \le 5$, then $n = 5$ is the only solution. Now we show that $n \ge 6$ cannot be a solution to the equation. Observe that $n \ge 6$ implies $n < 2^{n-3}$, thus

$$\frac{1+n}{2^n} < \frac{1+2^{n-3}}{2^n} = \frac{1}{2^n} + \frac{1}{8} < \frac{1}{16} + \frac{1}{8} = \frac{3}{16}.$$

So there are $\boxed{5}$ coins.

Exercises for Section 8.3

8.3.1 The only way that more than four can show ⊡ is if all 5 dice show ⊡, and the probability of that happening is $\frac{1}{6^5}$. Thus the answer is $1 - \frac{1}{6^5} = \boxed{\frac{7775}{7776}}$.

8.3.2 There are $2^6 = 64$ possible outcomes, since each of the 6 coins has 2 possibilities. If we do not get at least 2 heads, then we get either no heads or one heads. There is only 1 way to get 0 heads, and $\binom{6}{1} = 6$ ways to get 1 head, so the probability of getting at most one head is $\frac{7}{64}$. Therefore, the probability of getting at least 2 heads is $1 - \frac{7}{64} = \boxed{\frac{57}{64}}$.

8.3.3 The number of ways to draw out 3 balls from 15 is $\binom{15}{3} = 455$. We can choose 2 black balls and 1 white ball in $\binom{8}{2}\binom{7}{1} = 196$ ways. We can pick 1 black ball and 2 white balls in $\binom{8}{1}\binom{7}{2} = 168$ ways. Therefore we have $196 + 168 = 364$ ways to satisfy the condition, so the answer is $\frac{364}{455} = \boxed{\frac{4}{5}}$.

Alternatively, we could solve this problem by computing the complementary probability. If we don't pick 2 of one color and 1 of the other, that means all 3 balls were of the same color. There are

$\binom{8}{3} = 56$ ways to choose 3 black balls and $\binom{7}{3} = 35$ ways to choose 3 white balls, so there are $56 + 35 = 91$ ways to choose 3 balls of the same color. Thus the probability of choosing 2 of one color and 1 of the other is $1 - \dfrac{91}{455} = \dfrac{364}{455} = \boxed{\dfrac{4}{5}}$.

8.3.4 We will instead find the probability that the two E's are next to each other. There are $\dfrac{7!}{2}$ arrangements of the word SIXTEEN. If we want to find the number of arrangements such that the E's are next to each other, we find the number of arrangements of the six-letter word SIXT(EE)N (where we treat the two E's as a block), which is 6!. So the probability that an arrangement of the word SIXTEEN has the two E's next to each other is $\dfrac{6!}{\frac{7!}{2}} = \dfrac{2}{7}$. So the probability that the two E's aren't next to each other is $1 - \dfrac{2}{7} = \boxed{\dfrac{5}{7}}$.

8.3.5 The number of ways to choose a committee of all boys or all girls is $2 \times \binom{10}{4} = 420$. The total number of committees is $\binom{20}{4} = 4845$. Thus the answer is $1 - \dfrac{420}{4845} = \dfrac{4425}{4845} = \boxed{\dfrac{295}{323}}$.

Exercises for Section 8.4

8.4.1 In all parts we use the fact that there are $2^{10} = 1024$ possible outcomes of the 10 coin flips.

(a) There are $\binom{10}{8} = \binom{10}{2} = 45$ ways to get exactly 8 heads, so the probability is $\dfrac{45}{2^{10}} = \boxed{\dfrac{45}{1024}}$.

(b) The number of ways to get 8, 9, or 10 heads is $\binom{10}{8} + \binom{10}{9} + \binom{10}{10} = 45 + 10 + 1 = 56$. So the probability is $\dfrac{56}{1024} = \boxed{\dfrac{7}{128}}$.

(c) The probability that we flip at least 6 heads is equal to the probability that we flip at least 6 tails, by symmetry. Let's call this probability p. The only other possibility is that we flip exactly 5 heads and 5 tails, for which the probability is $\dfrac{\binom{10}{5}}{2^{10}} = \dfrac{252}{1024} = \dfrac{63}{256}$. Therefore, $\dfrac{63}{256} + 2p = 1$, giving

$$p = \dfrac{1}{2}\left(1 - \dfrac{63}{256}\right) = \boxed{\dfrac{193}{512}}.$$

8.4.2

(a) The chances of getting an odd or even number are equal, so there are $2^5 = 32$ equally likely outcomes: each die can be odd or even. If we want to get exactly 4 of 5 the rolls to be odd, the probability is $\dfrac{\binom{5}{4}}{2^5} = \boxed{\dfrac{5}{32}}$.

(b) The number of possible rolls of 5 dice is 6^5, since there are 6 possibilities for each of the 5 dice. Now we count the number of ways to get a ⚀ or a ⚁ in exactly 3 of the 5 rolls. First, we pick which 3 of the 5 rolls are ⚀ or ⚁: we can do that in $\binom{5}{3}$ ways. Now for each of these 3 rolls, there are 2 choices, and for each of the other 2 rolls, there are 4 choices. Thus the probability is

$$\frac{\binom{5}{3}2^3 4^2}{6^5} = \boxed{\frac{40}{243}}.$$

(c) The number of ways to roll exactly 2 ⚅'s is $\binom{5}{2}5^3$, since there are $\binom{5}{2}$ choices for which of the two dice are ⚅, and there are 5 choices for each of the other 3 dice. Similarly, the number of ways to roll exactly 1 ⚅ is $\binom{5}{1}5^4$, and the number of ways to roll no ⚅'s is $\binom{5}{0}5^5$. So the probability is

$$\frac{\binom{5}{2}5^3 + \binom{5}{1}5^4 + \binom{5}{0}5^5}{6^5} = \boxed{\frac{625}{648}}.$$

8.4.3 There are 30 days in June. The probability that it rains on exactly 0, 1, or 2 days is

$$\binom{30}{0}\left(\frac{1}{10}\right)^0\left(\frac{9}{10}\right)^{30} + \binom{30}{1}\left(\frac{1}{10}\right)^1\left(\frac{9}{10}\right)^{29} + \binom{30}{2}\left(\frac{1}{10}\right)^2\left(\frac{9}{10}\right)^{28} \approx \boxed{0.411}.$$

8.4.4 When Biff scores, the Grunters are more likely to win, because scoring improves their likelihood of winning. However, when Biff doesn't score, the Grunters are less likely to win. So we cannot make any conclusions since the two given probabilities are not independent of each other.

8.4.5 The probability that the Grunters win a 5-game series is

$$\binom{2}{0}\left(\frac{3}{4}\right)^3\left(\frac{1}{4}\right)^0 + \binom{3}{1}\left(\frac{3}{4}\right)^3\left(\frac{1}{4}\right)^1 + \binom{4}{2}\left(\frac{3}{4}\right)^3\left(\frac{1}{4}\right)^2 = \frac{459}{512} \approx 89.6\%,$$

and the probability that the Grunters win a 9-game series is

$$\binom{4}{0}\left(\frac{3}{4}\right)^5\left(\frac{1}{4}\right)^0 + \binom{5}{1}\left(\frac{3}{4}\right)^5\left(\frac{1}{4}\right)^1 + \binom{6}{2}\left(\frac{3}{4}\right)^5\left(\frac{1}{4}\right)^2 + \binom{7}{3}\left(\frac{3}{4}\right)^5\left(\frac{1}{4}\right)^3 + \binom{8}{4}\left(\frac{3}{4}\right)^5\left(\frac{1}{4}\right)^4$$

$$= \frac{249318}{262144} \approx 95.1\%.$$

Note that each term in each of the above expressions represents the probability that the Grunters win the series in some particular number of games. The coefficient of each term is the number of choices for the games which the Grunters lose.

So, indeed,

$$P(\text{Grunters win a single game}) < P(\text{Grunters win a 5-game series})$$
$$< P(\text{Grunters win a 7-game series})$$
$$< P(\text{Grunters win a 9-game series}).$$

In general, the longer the series, the more likely that the better team will win.

Exercises for Section 8.5

8.5.1

(a) The probability that the first card is a $\heartsuit$ is $\frac{1}{4}$. The second card then has a probability of $\frac{13}{51}$ of being

$\clubsuit$. So the answer is $\frac{1}{4} \times \frac{13}{51} = \boxed{\frac{13}{204}}$.

(b) The probability that the first card is a 6 is $\frac{1}{13}$. The probability the second card is a Queen is $\frac{4}{51}$.

The answer is then $\frac{1}{13} \times \frac{4}{51} = \boxed{\frac{4}{663}}$.

(c) We have two cases because if the first card is a King, it could be a $\heartsuit$ or not be a $\heartsuit$.

There is a $\frac{1}{52}$ chance that the King of $\heartsuit$ is drawn first, and a $\frac{12}{51} = \frac{4}{17}$ chance that the second card drawn is one of the twelve remaining $\heartsuit$, which gives a probability of $\frac{1}{52} \times \frac{4}{17} = \frac{1}{221}$ chance that this occurs.

There is a $\frac{3}{52}$ chance that a non-$\heartsuit$ King is drawn first, and a $\frac{13}{51}$ chance that a $\heartsuit$ is drawn second, giving a $\frac{3}{52} \times \frac{13}{51} = \frac{1}{68}$ chance that this occurs.

So the probability that one of these two cases happens is $\frac{1}{221} + \frac{1}{68} = \boxed{\frac{1}{52}}$. Note that this probability is the same as the probability of drawing the King of $\heartsuit$ on one card—can you determine why this is so?

(d) We have two cases because if the first card is a $\diamond$, it could be a Ace or not be an Ace.

There is a $\frac{1}{52}$ chance that the Ace of $\diamond$ is drawn first, and a $\frac{3}{51} = \frac{1}{17}$ chance that the second card drawn is one of the three remaining Aces, which gives a probability of $\frac{1}{52} \times \frac{1}{17} = \frac{1}{884}$ chance that this occurs.

There is a $\frac{12}{52} = \frac{3}{13}$ chance that a $\diamond$ other than the Ace is drawn first, and a $\frac{4}{51}$ chance that an Ace is drawn second, giving a $\frac{3}{13} \times \frac{4}{51} = \frac{4}{221}$ chance that this occurs.

So the probability that one of these two cases happens is $\frac{1}{884} + \frac{4}{221} = \boxed{\frac{1}{52}}$. As in part (c), we note that this probability is the same as the probability of drawing the Ace of $\diamond$ on one card—again, can you determine why this is so?

8.5.2

(a) The probability is $\frac{4}{52} \times \frac{4}{51} \times \frac{4}{50} = \boxed{\frac{8}{16575}}$.

(b) The probability is $\dfrac{13}{52} \times \dfrac{12}{51} \times \dfrac{11}{50} = \boxed{\dfrac{11}{850}}$.

(c) There are 4 exclusive cases:

Case 1: first card not a ♣ and second card not a 2. There are 3 cards that are 4's but not a ♣, so the probability for the first card is $\dfrac{3}{52}$. Next, there are 12 ♣s remaining that aren't a 2, so the probability for the second card is $\dfrac{12}{51}$. Finally, there are four 2's remaining, so the probability for the third card is $\dfrac{4}{50}$. Hence, this case gives a probability of $\dfrac{3}{52} \times \dfrac{12}{51} \times \dfrac{4}{50} = \dfrac{144}{132600}$. (We leave the fraction in these terms rather than reducing because we know that we're going to need to add fractions later.)

Case 2: first card not a ♣ and second card the 2♣. There are 3 cards that are 4's but not a ♣, so the probability for the first card is $\dfrac{3}{52}$. Next, there is only one 2♣, so the probability for the second card is $\dfrac{1}{51}$. Finally, there are three 2's remaining, so the probability for the third card is $\dfrac{3}{50}$. Hence, this case gives a probability of $\dfrac{3}{52} \times \dfrac{1}{51} \times \dfrac{3}{50} = \dfrac{9}{132600}$.

Case 3: first card the 4♣ and second card not a 2. There is only one 4♣, so the probability for the first card is $\dfrac{1}{52}$. Next, there are 11 ♣s remaining that aren't a 2, so the probability for the second card is $\dfrac{11}{51}$. Finally, there are four 2's remaining, so the probability for the third card is $\dfrac{4}{50}$. Hence, this case gives a probability of $\dfrac{1}{52} \times \dfrac{11}{51} \times \dfrac{4}{50} = \dfrac{44}{132600}$.

Case 4: first card the 4♣ and second card the 2♣. There is only one 4♣, so the probability for the first card is $\dfrac{1}{52}$. Next, there is only one 2♣, so the probability for the second card is $\dfrac{1}{51}$. Finally, there are three 2's remaining, so the probability for the third card is $\dfrac{3}{50}$. Hence, this case gives a probability of $\dfrac{1}{52} \times \dfrac{1}{51} \times \dfrac{3}{50} = \dfrac{3}{132600}$.

So the overall probability is $\dfrac{144 + 9 + 44 + 3}{132600} = \dfrac{200}{132600} = \boxed{\dfrac{1}{663}}$.

8.5.3 The probability that the first is red is $\dfrac{3}{8}$. Now with 7 remaining, the probability that the second is white is $\dfrac{5}{7}$. The answer is $\dfrac{3}{8} \times \dfrac{5}{7} = \boxed{\dfrac{15}{56}}$.

8.5.4 We can have all red, all white, or all blue. Thus the answer is

$$P(\text{all red}) + P(\text{all white}) + P(\text{all blue}) = \left(\dfrac{4}{15} \times \dfrac{3}{14} \times \dfrac{2}{13}\right) + \left(\dfrac{5}{15} \times \dfrac{4}{14} \times \dfrac{3}{13}\right) + \left(\dfrac{6}{15} \times \dfrac{5}{14} \times \dfrac{4}{13}\right) = \boxed{\dfrac{34}{455}}.$$

8.5.5 We can get the second marble to be yellow in two ways: either a white from A (with probability 3/7) then a yellow from B (with probability 6/10), or a black from A (with probability 4/7) then a yellow from C (with probability 2/7). Thus, the probability is

$$\left(\frac{3}{7} \times \frac{6}{10}\right) + \left(\frac{4}{7} \times \frac{2}{7}\right) = \boxed{\frac{103}{245}}.$$

8.5.6 The probability that it rains and Sheila attends is $(0.4)(0.2) = 0.08$. The probability that it doesn't rain and Sheila attends is $(0.6)(0.8) = 0.48$. So the overall probability that Sheila attends is $0.08 + 0.48 = \boxed{0.56 = 56\%}$.

Review Problems

8.17 There are 4 divisors of 6, namely $1, 2, 3, 6$. So the answer is $\dfrac{4}{6} = \boxed{\dfrac{2}{3}}$.

8.18 There are $\binom{10}{1}$ ways to roll exactly one $\boxed{\cdot}$ out of 10 dice. The probability of any one of these occurring is $\left(\frac{1}{6}\right)^1 \left(\frac{5}{6}\right)^9$. So the overall probability is

$$\binom{10}{1}\left(\frac{1}{6}\right)^1\left(\frac{5}{6}\right)^9 = \boxed{\frac{10 \times 5^9}{6^{10}}} \approx 0.323.$$

8.19 There are $\binom{6}{3}$ ways for 3 of the dice to show even numbers and 3 of them to show odd numbers. Each roll is even with probability $\frac{1}{2}$ and odd with probability $\frac{1}{2}$, so each arrangement of 3 odd numbers and 3 even numbers occurs with probability $\left(\frac{1}{2}\right)^6$. Thus, the probability that 3 dice out of 6 show even numbers is

$$\binom{6}{3}\frac{1}{2^6} = \boxed{\frac{5}{16}}.$$

8.20 There are $\binom{n}{k}$ ways for k of the coins to land on heads and $n - k$ of the coins to land on tails. Each of these sequences has a probability $p^k(1 - p)^{n-k}$, since heads are flipped k times and tails are flipped $n - k$. Thus, the probability that exactly k of the coins come up heads is $\boxed{\binom{n}{k}p^k(1 - p)^{n-k}}$.

8.21

(a) There are 12 face cards, so there are $\binom{12}{2}$ ways to choose 2 face cards (without regard to order). There are $\binom{52}{2}$ ways to choose any 2 cards (without regard to order). So the answer is

$$\frac{\binom{12}{2}}{\binom{52}{2}} = \boxed{\frac{11}{221}}.$$

(b) There are two cases.

Case 1: The first card is a ♡ but not a 10. The probability of the first card satisfying this is $\frac{12}{52}$, and then the probability that the second card is a 10 is $\frac{4}{51}$.

Case 2: The first card is the 10♡. The probability of the first card being the 10♡ is $\frac{1}{52}$, and then the probability that the second card is a 10 is $\frac{3}{51}$.

We then add the probability of the two cases (since they are exclusive) to get

$$\frac{12}{52} \times \frac{4}{51} + \frac{1}{52} \times \frac{3}{51} = \boxed{\frac{1}{52}}.$$

(c) There are two cases that we have to consider.

Case 1: The first card is one of 2,3,4,5,7,8,9,10. There are 32 such cards, so this occurs with probability $\frac{32}{52}$. For any of these cards, there are 4 cards left in the deck such that the cards sum to 12, so the probability of drawing one is $\frac{4}{51}$. Thus, the probability that this case occurs is $\frac{32}{52} \times \frac{4}{51} = \frac{32}{663}$.

Case 2: The first card is a 6. There are 4 of these, so this occurs with probability $\frac{4}{52}$. Now we need to draw another 6. There are only 3 left in the deck, so the probability of drawing one is $\frac{3}{51}$. Thus, the probability that this case occurs is $\frac{4}{52} \times \frac{3}{51} = \frac{3}{663}$.

Therefore the overall probability is $\frac{32}{663} + \frac{3}{663} = \boxed{\frac{35}{663}}$.

8.22 We can find the probability they are all same color, then subtract that from 1. There are 26 cards of each color, so 3 of them can be selected in $\binom{26}{3}$ ways, and of course there are 2 colors. So the answer is

$$1 - 2\frac{\binom{26}{3}}{\binom{52}{3}} = \boxed{\frac{13}{17}}.$$

8.23 The probability is given by

$$P(\text{green ball}) = (P(\text{Box I}) \times P(\text{green from Box I})) + (P(\text{Box II}) \times P(\text{green from Box II}))$$
$$+ (P(\text{Box III}) \times P(\text{green from Box III}))$$
$$= \left(\frac{1}{3} \times \frac{4}{12}\right) + \left(\frac{1}{3} \times \frac{4}{6}\right) + \left(\frac{1}{3} \times \frac{4}{6}\right)$$
$$= \frac{1}{9} + \frac{2}{9} + \frac{2}{9} = \boxed{\frac{5}{9}}.$$

Note that it is incorrect to simply say: "There are 12 green balls and 12 red balls total, so the probability is $\frac{1}{2}$." This is because not every ball is equally likely to be chosen: the balls in containers II and III are more likely than the balls in container I.

8.24

(a) We could have either two greens or two reds. The probability of drawing two greens is $\left(\frac{6}{10}\right)^2 = \frac{9}{25}$. The probability of drawing two reds is $\left(\frac{4}{10}\right)^2 = \frac{4}{25}$. So the answer is $\frac{9}{25} + \frac{4}{25} = \boxed{\frac{13}{25}}$.

(b) Again, we could have either two greens or two reds. The probability of drawing two greens is $\frac{6}{10} \times \frac{5}{9} = \frac{1}{3}$. The probability of drawing two reds is $\frac{4}{10} \times \frac{3}{9} = \frac{2}{15}$. So the answer is $\frac{1}{3} + \frac{2}{15} = \boxed{\frac{7}{15}}$.

8.25 We can break up the probability into cases as follows:

$$P(\text{more heads on pennies than nickels}) = P(3 \text{ heads on pennies})$$
$$+ P(2 \text{ heads on pennies}) \times P(0 \text{ or } 1 \text{ heads on nickels})$$
$$+ P(1 \text{ head on pennies}) \times P(0 \text{ heads on nickels}).$$

So the probability is

$$\left(\frac{1}{2}\right)^3 + 3\left(\frac{1}{2}\right)^3\left(\frac{3}{4}\right) + 3\left(\frac{1}{2}\right)^3\left(\frac{1}{4}\right) = \frac{1}{8} + \frac{9}{32} + \frac{3}{32} = \boxed{\frac{1}{2}}.$$

As a challenge, try to prove that this probability is always $\frac{1}{2}$, as long as we have one more penny than nickel.

8.26 The probability that the MegaBall matches is $\frac{1}{27}$. The probability that the 5 WinnerBalls match is $\frac{1}{\binom{44}{5}}$. So my chances of winning are $\left(\frac{1}{27}\right) \times \left(\frac{1}{\binom{44}{5}}\right) = \boxed{\frac{1}{29,322,216}}$.

8.27

(a) The second die has a $\boxed{\frac{1}{6}}$ probability of matching the first.

(b) The probability that doubles are rolled 3 times in a turn is $\left(\frac{1}{6}\right)^3 = \frac{1}{216}$. Therefore the chance of not going to jail is $1 - \frac{1}{216} = \boxed{\frac{215}{216}}$.

Challenge Problems

8.28 There are two cases.

Case 1: the president is a boy. There are 7 choices for the president. Then we need to count the number of ways to choose a committee with more boys than girls. There are $\binom{6}{3}$ ways to choose 3 boys for the committee, and there are $9 \times \binom{6}{2}$ ways to choose a girl and 2 boys. So there are a total of

$$7\left(\binom{6}{3} + 9\binom{6}{2}\right) = 7(20 + 135) = 1085$$

choices in this case.

Case 2: the president is a girl. There are 9 choices for the president. Then we need to count the number of ways to choose a committee with more girls than boys. There are $\binom{8}{3}$ ways to choose 3 girls for the committee, and there are $7 \times \binom{8}{2}$ ways to choose a boy and 2 girls. So there are a total of

$$9\left(\binom{8}{3} + 7\binom{8}{2}\right) = 9(56 + 196) = 2268$$

choices in this case.

So there are $1085 + 2268 = 3353$ successful ways to choose the president and committee such that the president's gender matches that of the majority of the committee. Overall, there are 16 choices for the president, then $\binom{15}{3}$ choices for the committee. So the desired probability is

$$\frac{3353}{16 \times \binom{15}{3}} = \frac{3353}{7280} = \boxed{\frac{479}{1040}}.$$

8.29 We consider as different cases the number of goals that Vanessa has when Richard wins. If Richard wins after Vanessa scores n goals, there are $\binom{4+n}{4}$ different sequences that Richard and Vanessa could have scored goals, because Richard must score the last goal, and there are $4 + n$ remaining places to put his other 4 scores. The probability that any one of these ways occurs is $(0.4)^5(0.6)^n$, which means Richard has a $\binom{4+n}{4}(0.4)^5(0.6)^n$ chance of winning by a score of 5-to-n. We sum up all these n from 0 to 4 to get Richard's total probability of winning, which is:

$$\binom{4}{4}(0.4)^5(0.6)^0 + \binom{5}{4}(0.4)^5(0.6)^1 + \binom{6}{4}(0.4)^5(0.6)^2 + \binom{7}{4}(0.4)^5(0.6)^3 + \binom{8}{4}(0.4)^5(0.6)^4 = \boxed{\frac{104128}{390625} \approx .267}$$

8.30 I lose if I fail to roll a ⚁ on any of my first six rolls. On each individual roll there is a $\frac{5}{6}$ chance that I fail to roll a ⚁, so the probability that I fail to roll a ⚁ in six rolls is $\left(\frac{5}{6}\right)^6$. Therefore the probability of the game ending in six rolls or less is

$$1 - \left(\frac{5}{6}\right)^6 = \frac{6^6 - 5^6}{6^6} = \boxed{\frac{31,031}{46,656}}.$$

Alternatively, we could calculate this probability directly. The probability that the game ends after 1 roll is $\frac{1}{6}$. The probability that it ends after 2 rolls is $\left(\frac{5}{6}\right) \times \left(\frac{1}{6}\right)$: I have to not roll a ⚁ on my first roll, then roll a ⚁ on my second roll. The probability that it ends after 3 rolls is $\left(\frac{5}{6}\right)^2 \times \left(\frac{1}{6}\right)$, by the same reasoning. The probability that it ends after 4 rolls is $\left(\frac{5}{6}\right)^3 \times \left(\frac{1}{6}\right)$, and so on. So the probability that the game ends on 6 rolls or less is

$$\frac{1}{6} + \left(\frac{5}{6} \times \frac{1}{6}\right) + \left(\left(\frac{5}{6}\right)^2 \times \frac{1}{6}\right) + \cdots + \left(\left(\frac{5}{6}\right)^5 \times \frac{1}{6}\right) = \frac{6^5 + 5(6^4) + (5^2)(6^3) + (5^3)(6^2) + (5^4)6 + 5^5}{6^6} = \boxed{\frac{31,031}{46,656}}.$$

8.31 Let p denote the probability that a ⚃ is rolled. Then we can see that $P(⚃) = P(⚄) = P(⚅) = 2p$ and $P(⚀) = P(⚁) = \frac{2}{3}p$. Since all of these probabilities must add to one, we get

$$1 = P(⚀) + P(⚁) + P(⚂) + P(⚃) + P(⚄) + P(⚅)$$
$$= \frac{2}{3}p + \frac{2}{3}p + 2p + 2p + 2p + p = \frac{25}{3}p.$$

Therefore, $p = \boxed{\dfrac{3}{25}}$.

8.32

(a) The only possibilities are $2, 2, 2$ and $1, 2, 3$. There is only 1 way to draw $2, 2, 2$, but there are $3! = 6$ ways to draw $1, 2, 3$. Thus there are $\boxed{7}$ ways.

(b) Only one of the 7 possible ways found in part (a) is $2, 2, 2$. So the answer is $\boxed{\dfrac{1}{7}}$. (This is an example of a type of problem called "conditional probability." This topic will be a major focus of the *Intermediate Counting & Probability* book.)

8.33 The probability that Wayne wins after $(2k - 1)$ flips is given by $\left(\dfrac{1}{2}\right)^{2k-1}$. This is because every flip must be a head until the very last flip. Thus, the probability he wins is the sum

$$\frac{1}{2} + \left(\frac{1}{2}\right)^3 + \left(\frac{1}{2}\right)^5 + \left(\frac{1}{2}\right)^7 + \cdots = \frac{1}{2}\left(1 + \frac{1}{4} + \left(\frac{1}{4}\right)^2 + \cdots\right) = \frac{\frac{1}{2}}{1 - \frac{1}{4}} = \boxed{\frac{2}{3}}.$$

Another way to look at this problem is as follows: Wayne wins on the first turn with probability $\frac{1}{2}$. Mario wins on his first turn with probability $\frac{1}{4}$. If they both flip heads, then the game essentially resets to the beginning. So Wayne is twice as likely to win as Mario. If Wayne's probability of winning is p, then Mario's probability of winning is $p/2$, and since $p + (p/2) = 1$, we get $p = \boxed{2/3}$.

8.34 Divide up the 52 cards into 4 hands. We will check the Aces one at a time to see if they are in the same hand as any other Ace. First we find out what hand the Ace of ♠ is in. Now we check the Ace of ♣. It cannot be in the same hand as the Ace of ♠. There are 39 cards not in the hand of the Ace of ♠ and 51 cards other than the Ace of ♠, so the Ace of ♣ has a $\dfrac{39}{51}$ chance of not being in the hand of the ♠. Similarly, the Ace of ♡ cannot be in the hand of the Ace of ♠ or ♣, so there is a $\dfrac{26}{50}$ chance of this happening, and a $\dfrac{13}{49}$ chance that the Ace of ♢ is not in the hands of any of the first three Aces. So the probability that all four Aces are in different hands is $\dfrac{39}{51} \times \dfrac{26}{50} \times \dfrac{13}{49} = \boxed{\dfrac{2197}{20825}}$.

8.35 The probability that no two have the same birthday is easier to calculate. We shall find that, and subtract it from 1 to get our answer. We will check the 100 people in order. The first person can have any birthday, the second person has a $\dfrac{364}{365}$ chance of having a different birthday from the first person, the third person has a $\frac{363}{365}$ chance of having a different birthday from the first two, and so on. Therefore the probability that everyone has a different birthday is $\dfrac{365 \times 364 \times \cdots \times 266}{365^{100}}$, and thus the probability that at least two have the same birthday is

$$1 - \frac{365 \times 364 \times \cdots \times 266}{365^{100}} = 1 - \frac{365!}{265!(365)^{100}} \approx .9999997\ldots$$

8.36 To have at least a $\frac{4}{5}$ probability of having a good picture, I need to have less than a $\frac{1}{5}$ probability of all of my pictures being bad. After taking n pictures, the chance that in each picture someone is looking away is given by $\left(\frac{3}{4}\right)^n$. We want to choose n large enough so that $\left(\frac{3}{4}\right)^n < \frac{1}{5}$. A little calculation shows that $\left(\frac{3}{4}\right)^5 \approx 0.237$ and $\left(\frac{3}{4}\right)^6 \approx 0.178$. So I will need to take $\boxed{6}$ pictures.

8.37

(a) Assuming that the odds are fair, horse A being 2-to-1 implies that the probability of its winning is $\frac{1}{3}$, and similarly horse B being 3-to-1 implies that the probability of its winning is $\frac{1}{4}$. Therefore, the probability that horse C wins is $1 - \frac{1}{3} - \frac{1}{4} = \frac{5}{12}$, so the fair odds on horse C should be $\frac{7}{12}$-to-$\frac{5}{12}$, or $\boxed{\text{7-to-5}}$. The "fair" betting system would then be to bet \$4 on horse A, \$3 on horse B, and \$5 on horse C—no matter which horse then wins, you would break even.

(b) There are many possible betting schemes. For example, we could bet \$15 on horse A and \$10 on each of horses B and C. Then, no matter which horse wins, we'll make money, as seen in the following chart:

Winning Horse	Money Won	Money Lost	Net Profit
A	\$30	\$20	\$10
B	\$30	\$25	\$5
C	\$40	\$25	\$15

As a challenge, see if you can find a system which guarantees the same amount of profit, no matter which horse wins.

CHAPTER **9**

_____**Think About It!**

Challenge Problems

9.8 Four socks are being drawn, and there are only three colors of socks. This means at least one pair of the same color sock is drawn: if the first 3 drawn were all different colors, then the fourth sock must match one of them. So the probability that a matching pair is drawn is $\boxed{1}$.

9.9 Since 11 is odd, for any series of 11 flips, there are either more heads than tails, or more tails than heads. Since the probabilities of getting heads or tails on an individual flip are equal, the probabilities of getting more heads than tails or more tails than heads are also equal, and they must be $\boxed{1/2}$.

9.10 When 37 is placed into a group, there are 49 remaining numbers left in the group, and 99 remaining integers. Since 89 is placed randomly as well, the chance that it is placed into same group as 37 is $\boxed{\dfrac{49}{99}}$.

9.11 By thinking about it a little bit, we can see that

$$P(\boxdot \text{ or } \boxed{\cdots}) = P(\boxed{\cdot\cdot} \text{ or } \boxed{\cdots}) = P(\boxed{\cdot\cdot} \text{ or } \boxed{\vdots}),$$

and since they have to sum to 1, we conclude that they each must be $\boxed{1/3}$. (The fact that $\boxed{\cdots}$ is four times as likely as $\boxed{\cdot\cdot}$ is irrelevant.)

9.12

(a) Let F be a full drive, H be a half-way drive, and M be a miss. The sequences that could result in the nail not going all the way through are (MMMM, HMMM, MHMM, MMHM, MMMH).

(b) For each sequence in part (a), each individual swing occurs with probability 1/4. So the probability of the sequence occurring is $(1/4)^4 = 1/256$.

(c) The probability that the nail is not driven in is the sum of the probabilities of the individual sequences, which is just $\boxed{5/256}$.

9.13

(a) Reverse the sequence of draws of coins. Now the question is asking what the probability is that the first coin pulled out is a penny. So it's just $\boxed{\dfrac{2}{13}}$.

(b)   The probability that any one coin selected is a nickel is $\frac{7}{13}$. If this is a nickel, then the probability that any other coin selected is a quarter is $\frac{4}{12}$, because there are 12 remaining coins. So the chance that the fourth coin is a nickel and the seventh is a quarter is $\frac{7}{13} \times \frac{4}{12} = \boxed{\frac{7}{39}}$.

9.14

(a)   Whichever Joe's the first friend goes to, each of the other four friends has a $\frac{1}{5}$ chance of going to the same one. So the probability that they all end up at the same one is $\left(\frac{1}{5}\right)^4 = \boxed{\frac{1}{625}}$.

(b)   After the first friend chooses his Joe's, the second one has a $\frac{4}{5}$ chance of going to a different Joe's from the first friend, the third friend has a $\frac{3}{5}$ chance of going to a different Joe's from the first two, the fourth friend has a $\frac{2}{5}$ chance of going to a different Joe's from the first three, and the fifth friend has a $\frac{1}{5}$ chance of going to a different Joe's from the first four. So the probability that they end up at different Joe's is $\frac{4 \times 3 \times 2 \times 1}{5^4} = \boxed{\frac{24}{625}}$.

9.15   We note that only the last three digits matter, so we can ignore the first 5 rolls. Clearly the number must be even, so the last digit must be 2, 4, or 6. At this point, it is easy to just list them:

Last digit	3-digit multiples of 8 with all digits between 1 and 6
2	112, 152, 232, 312, 352, 432, 512, 552, 632
4	144, 224, 264, 344, 424, 464, 544, 624, 664
6	136, 216, 256, 336, 416, 456, 536, 616, 656

There are 27 successful outcomes, and $6^3 = 216$ possible outcomes for the last three dice, so the probability is $\frac{27}{216} = \boxed{\frac{1}{8}}$.

There are quicker ways to solve the problem using a more advanced knowledge of number theory—can you see how?

9.16   Let the probability that Timmy beats the first parent he plays be a, and the probability that Timmy beats the other parent b. There are two ways that Timmy can win: he can win the first two games, or he can lose the first and win the next two.

Case 1: Timmy wins the first two. The probability that he beats the first parent is a, and the probability he beats the second parent is b, so Timmy has an ab chance of doing this.

Case 2: Timmy loses the first one and wins the next two. The probability that he loses to the first parent is $1 - a$, the probability that he beats the other parent is b, and the probability that he beats the first parent in the third game is a. The probability that Timmy does this is $(1 - a)ba$.

Summing these two probabilities together, Timmy has a $(1 - a)ba + ab = (2 - a)ab$ chance of going to the concert. So if Timmy plays his mother first, he has a $(2 - m)mf$ probability of going to the concert, and if he plays his father first, he has a $(2 - f)mf$ chance of going to the concert. Since $m < f$, we have $2 - m > 2 - f$, which means $(2 - m)mf > (2 - f)mf$. So Timmy has a better chance of going to the concert if he plays his $\boxed{\text{mother}}$ first.

Intuitively, this answer should make sense. Timmy has only one chance to beat his second opponent—he must win that game in order to win the match. So he should play the weaker opponent in that game. He only has to beat his other opponent in one of that player's two games, so that's when he should play the stronger opponent.

9.17 (Note: we abbreviate the names of the 3 players as X, Y, and Z.) When X takes his first shot, three things can happen: he hits Y, he hits Z, or he misses. We will examine each of these three possibilities.

Case 1: X hits Y. This is really bad for X, since this means that Z will go next and hit X. X has no chance of winning.

Case 2: X hits Z. This means Y and X will face off in a duel against each other, with Y shooting first. X has at least a 2/3 chance of losing this duel, so less than a 1/3 chance of winning.

Case 3: X misses. Y must shoot at Z, because if Y doesn't hit Z, Z will hit Y (Z will always hit Y because Z would rather be shot at by X than Y).

    Case 3a: If Y hits Z, then X and Y are in a duel with X shooting first. X has at least a 1/3 chance of winning this.

    Case 3b: If Y misses Z, then Z hits Y. Now X has a shot at Z before Z hits X. X has a 1/3 chance of winning this.

If X hits someone on its first shot, X has less than a 1/3 chance of winning, while if X doesn't hit someone on the shot, X has greater than a 1/3 chance of winning. This means X's best strategy is to miss intentionally!

Now we must find the probability of X winning. Since Y has a $\dfrac{2}{3}$ chance of hitting Z,

$$P(\text{X wins}) = \frac{2}{3} \times P(\text{X wins in Case 3a}) + \frac{1}{3} \times P(\text{X wins in Case 3b}).$$

Case 3a: X and Y are in a duel with X shooting first. X has a $\dfrac{1}{3}$ chance of hitting Y on the first shot, a $\dfrac{2}{3} \times \dfrac{1}{3} \times \dfrac{1}{3}$ chance of hitting Y on the third shot of the duel (X and Y both miss and then X hits Y), a $\dfrac{2}{3} \times \dfrac{1}{3} \times \dfrac{2}{3} \times \dfrac{1}{3} \times \dfrac{1}{3}$ of hitting Y on the fifth shot of the duel, etc. So the probability of X winning in this case is

$$\frac{1}{3} + \left(\frac{2}{3} \times \frac{1}{3} \times \frac{1}{3}\right) + \left(\frac{2}{3} \times \frac{1}{3} \times \frac{2}{3} \times \frac{1}{3} \times \frac{1}{3}\right) + \cdots = \frac{1}{3} \times \left(1 + \frac{2}{9} + \left(\frac{2}{9}\right)^2 + \cdots\right) = \frac{\frac{1}{3}}{1 - \frac{2}{9}} = \frac{3}{7}.$$

Case 3b: If Y misses Z on its first shot, Z will hit Y. X must shoot at Z, and has a $\dfrac{1}{3}$ chance of hitting Z and winning. If X misses, he loses because Z will hit X immediately. So the probability of X winning in this case is $\frac{1}{3}$.

So

$$P(X \text{ wins}) = \frac{2}{3} \times P(X \text{ wins in Case 3a}) + \frac{1}{3} \times P(X \text{ wins in Case 3b}) = \frac{2}{3} \times \frac{3}{7} + \frac{1}{3} \times \frac{1}{3} = \boxed{\frac{25}{63}}.$$

9.18 We can see that the height of the chosen rectangle is irrelevant, so we can assume that we start with just a single row of 2005 squares, with the middle one shaded. To choose a smaller rectangle within this square, we must choose the leftmost square and the rightmost square: these can be the same (producing a 1×1 rectangle) in 2005 ways, and different in $\binom{2005}{2}$ ways, for a total of

$$2005 + \binom{2005}{2} = 2005 + \frac{(2005)(2004)}{2} = \frac{(2)(2005) + (2005)(2004)}{2} = \frac{(2005)(2006)}{2} = \binom{2006}{2}$$

ways. To choose a rectangle which contains the shaded square, the left end must be one of the 1003 squares to the left of (and including) the shaded square, and the right end must be one of the 1003 squares to the right of (and including) the shaded square, so there are $(1003)^2$ ways we can choose such a rectangle. Therefore the probability is

$$\frac{(1003)^2}{\binom{2006}{2}} = \frac{2(1003)(1003)}{(2006)(2005)} = \boxed{\frac{1003}{2005}}.$$

CHAPTER

10

_____ **Geometric Probability**

Exercises for Section 10.2

10.2.1 Let C be the midpoint of $\overline{AB}$, and let D, E be midpoints of $\overline{AC}, \overline{CB}$ respectively. We can see that P is closer to the midpoint of $\overline{AB}$ than to A or B if and only if P lies on segment $\overline{DE}$. The probability of this is

$$\frac{\text{length of } \overline{DE}}{\text{length of } \overline{AB}} = \frac{5}{10} = \boxed{\frac{1}{2}}.$$

$$\overset{10}{\underset{A \; \; 2.5 \; \; D \; \; 2.5 \; \; C \; \; 2.5 \; \; E \; \; 2.5 \; \; B}{\rule{7cm}{0.4pt}}}$$

10.2.2 We have $x^2 < 2$ if and only if $-\sqrt{2} < x < \sqrt{2}$. Therefore the answer is

$$\frac{\text{Length of segment with } -\sqrt{2} < x < \sqrt{2}}{\text{Length of segment with } -2 \le x \le 5} = \frac{\sqrt{2} - (-\sqrt{2})}{5 - (-2)} = \boxed{\frac{2\sqrt{2}}{7}}.$$

10.2.3 Note that $y - \lfloor y \rfloor$ is the fractional part of y. Thus $0 \le y - \lfloor y \rfloor < 1$. We have $y - \lfloor y \rfloor \ge 1/3$ if and only if the fractional part of y is at least $1/3$. The probability is

$$\frac{\text{Length of segment with } 1/3 \le y < 1}{\text{Length of segment with } 0 \le y < 1} = \frac{1 - 1/3}{1 - 0} = \boxed{\frac{2}{3}}.$$

10.2.4 Let $\overline{CD}$ be the line segment $0 \le x \le 6$ on the real line. Then the condition is that $x < (6-x)^2$. This means that $x^2 - 13x + 36 > 0$, or $(x-4)(x-9) > 0$. Hence we must have $x < 4$ or $x > 9$; the latter condition makes no sense given that $\overline{CD}$ consists only of the points from 0 to 6. Therefore our probability is:

$$\frac{\text{length of segment with } 0 \le x < 4}{\text{length of } \overline{CD}} = \frac{4}{6} = \boxed{\frac{2}{3}}.$$

10.2.5 The coordinates of P are $(x, x+1)$ for some x in the interval $(0, 3)$. The triangle with vertices $(0, 0)$, $(3, 0)$, and $(x, x+1)$ has base 3 and height $x + 1$, so its area is $\frac{3(x+1)}{2} = \frac{3x+3}{2}$. We must have $\frac{3x+3}{2} > 2$, therefore $x > \frac{1}{3}$. So the probability is $\dfrac{3 - \frac{1}{3}}{3 - 0} = \boxed{\frac{8}{9}}$.

Exercises for Section 10.3

10.3.1 For each of the following problems, since the area of the square is 1, the probability is just

$$P(\text{event}) = \frac{\text{area of successful outcome region}}{\text{area of possible outcome region}} = \text{area of successful outcome region}.$$

So we simply need to find the area of the region described by the given conditions. In all of the following, the shaded portions of the diagrams will indicate our successful region.

(a) This is a right triangular region with legs 1/2. The answer is $\boxed{\dfrac{1}{8}}$.

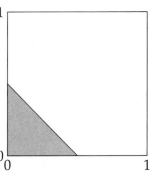

(b) This is a trapezoidal region with bases 1/2 and 1 and height 1. The answer is $\boxed{\dfrac{3}{4}}$.

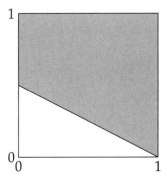

(c) This is a strip region. We can subtract the areas of the two triangles from that of the square. The two triangles put together form a 0.8×0.8 square. The answer is $1 - 0.8^2 = 0.36 = \boxed{\dfrac{9}{25}}$.

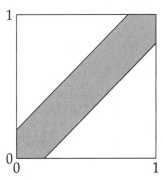

(d) This is a quarter of the unit circle centered at the origin. The answer is $\boxed{\dfrac{\pi}{4}}$.

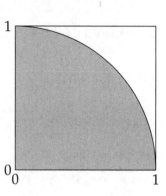

(e) This is a circle of radius 1/2 inscribed in the unit square. The answer is $\boxed{\dfrac{\pi}{4}}$.

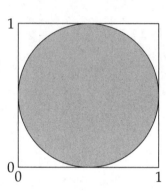

(f) This region is the unit square minus a quarter unit circle. The answer is $\boxed{1 - \dfrac{\pi}{4}}$.

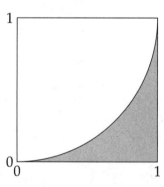

10.3.2 Since the tiles' pattern repeats over the entire floor, we can look at just one tile to find the probability. The denominator of our probability is the area of one tile, namely 100. The penny lies entirely within the tile if and only if the penny's center falls in the center 8×8 square of the tile, a region with area 64. So the answer is $\dfrac{64}{100} = \boxed{\dfrac{16}{25}}$.

10.3.3 Let my arrival time be x and my friend's arrival time be y. We would meet if and only if $x - 20 \le y \le x + 15$, because I will only wait 15 minutes for him, and he will wait 20 minutes for me. This region is a strip, bordered by two lines parallel to $y = x$, and its area can be found by subtracting the two corner triangles from square, which is the complement of its area. The answer is

$$\frac{60^2 - \frac{40^2}{2} - \frac{45^2}{2}}{60^2} = \boxed{\frac{143}{288}}.$$

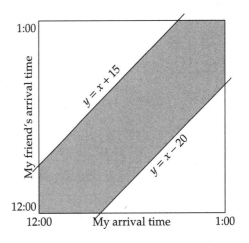

10.3.4 Let x denote the time when Maryanne's mail arrives and y denote the time when she goes to check on it. So we have two random variables $1 \le x \le 3$ and $2 \le y \le 3$. This makes the "possible" region be a rectangular region of area 2. The region that represents the mail having already been delivered when she goes to check on it is given by $y \ge x$. This is a trapezoid of area $\frac{3}{2}$. So the probability is $\dfrac{\frac{3}{2}}{2} = \boxed{\frac{3}{4}}$.

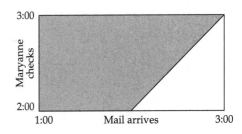

There is also a quick, "think about it" solution. There is a $\dfrac{1}{2}$ probability that the mail arrives between 1:00 and 2:00, in which case Maryanne will always be able to pick it up. Otherwise, there's a $\dfrac{1}{2}$ probability that the mail arrives between 2:00 and 3:00. In this case, it's equally likely that the mail or Maryanne will arrive first, so there's a $\dfrac{1}{2} \times \dfrac{1}{2} = \dfrac{1}{4}$ probability that the mail arrives after 2:00 and Maryanne is able to pick it up. Therefore, the total probability is $\dfrac{1}{2} + \dfrac{1}{4} = \boxed{\dfrac{3}{4}}$.

Review Problems

10.7 The interval for which $x^2 \le \dfrac{1}{2}$ is the interval in which $-\dfrac{\sqrt{2}}{2} \le x \le \dfrac{\sqrt{2}}{2}$. This segment has length $\sqrt{2}$. The probability that $x^2 \le \dfrac{1}{2}$ is then $\dfrac{\sqrt{2}}{2}$. We want $x^2 > \dfrac{1}{2}$, so the answer is $\boxed{1 - \dfrac{\sqrt{2}}{2}}$.

10.8 First fix point P, since we're only looking at the relative position of the two points. Then Q can fall anywhere within 60 degrees left or right of P, which is an arc of 120 degrees. Therefore the probability is $\dfrac{120}{360} = \boxed{\dfrac{1}{3}}$.

10.9 The y-coordinate of R can be anywhere from 0 to 10. We want it between 7 and 10. Therefore, the probability is $\dfrac{10-7}{10-0} = \boxed{\dfrac{3}{10}}$.

10.10 By symmetry, any point is just as likely to be closer to A than to any of the other 7 vertices. So the answer is $\boxed{1/8}$.

10.11 For each of the following problems, we need to find the area of the region described by the given condition. We then divide that area by 2 to get the probability, since we are working in a 1×2 rectangle.

(a) The region is a trapezoid with area $\dfrac{3}{2}$. Answer is $\boxed{\dfrac{3}{4}}$.

(b) The region is a triangle with area $\dfrac{1}{2}$. Answer is $\boxed{\dfrac{1}{4}}$.

(c) The region is made up of points that lie outside the quarter unit circle centered at the origin. The area is $2 - \dfrac{\pi}{4}$, so the answer is $\boxed{1 - \dfrac{\pi}{8}}$.

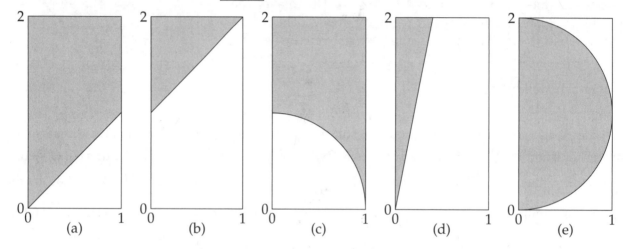

(d) The region is a triangle of area $\dfrac{2}{5}$. The answer is $\boxed{\dfrac{1}{5}}$.

(e) We are looking at $x^2 + (y-1)^2 < 1$. This is the semicircle of radius 1 centered at $(0,1)$. Its area is $\dfrac{\pi}{2}$, and the answer is $\boxed{\dfrac{\pi}{4}}$.

10.12   The points that lie on the perpendicular bisector l of $(0,0)$ and $(3,3)$ are equidistant to the two points. So all of the points below the line l are points closer to $(0,0)$ than to $(3,3)$. So these are points such that $x + y < 3$. This is all the points inside the square except for the right triangle bordered by $(2,1)$, $(1,2)$, and $(2,2)$, which has area $\dfrac{1}{2}$, which means that the successful region has area $4 - \dfrac{1}{2} = \dfrac{7}{2}$. Thus the answer is $\dfrac{\frac{7}{2}}{4} = \boxed{\dfrac{7}{8}}$.

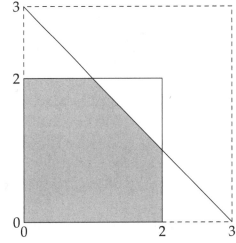

10.13   Frankie will take the express if she arrives during one of the following time intervals in any hour: between :00 past and :05 past, between :15 past and :25 past, or between :30 past and :45 past. Since we are looking at a total time interval of an hour (which is 60 minutes), the probability is $\dfrac{5 + 10 + 15}{60} = \boxed{\dfrac{1}{2}}$.

10.14   The product ab is positive if and only if a and b are both positive or both negative. These regions are shaded in the picture to the right. The two shaded rectangles have area $6 + 4 = 10$, and the total possible area is $6 \times 4 = 24$, so the probability is $\dfrac{10}{24} = \boxed{\dfrac{5}{12}}$.

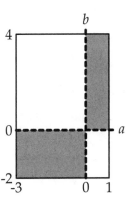

Challenge Problems

10.15   Since x, y, z are randomly and independently placed, any ordering for them is equally likely. There are $3! = 6$ orderings for 3 random numbers, so the chance that it is any one particular order (in this case $x \le y \le z$) is $\boxed{\dfrac{1}{6}}$.

10.16   Three lengths a, b, c can be the sides of the triangle if and only if they satisfy the Triangle Inequality in every possible way; that is, if $a + b > c$, $b + c > a$, and $c + a > b$. Suppose that our line segment is the unit interval, and we are breaking it at points x and y with $0 < x < y < 1$. Then the three resulting segments will form a triangle if and only if $x < \frac{1}{2}$, $y > \frac{1}{2}$, and $y - x < \frac{1}{2}$. So the "possible" region is the triangle with vertices $(0,0)$, $(0,1)$, and $(1,1)$, and the "successful" region is as shaded in the diagram on the right. Therefore the probability of success is $\dfrac{\frac{1}{8}}{\frac{1}{2}} = \boxed{\dfrac{1}{4}}$.

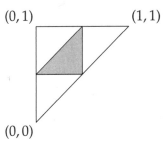

10.17 Since the triangles ABC and PBC have the same base, we simply need the
height of PBC to be less than half of the height of ABC. This means that the distance
from P to BC must be less than half the distance from A to BC. If we let D and E be
the midpoints of sides AB and AC respectively, then we must have P inside of the
trapezoid $DECB$, as shown in the diagram. The triangle ADE is similar to ABC,
with side lengths in a $1 : 2$ ratio. So the area of ADE is $1/4$ that of ABC, meaning
that the area of $DECB$ is $3/4$ that of ABC, giving a probability of $\boxed{3/4}$.

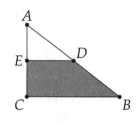

10.18 Locate point D_0 on AC such that $BD_0 = \sqrt{2}$. Point P will satisfy the
condition of the problem if and only if P lies in triangle BCD_0, which has area $\dfrac{1}{2}$.

ABC has area $\dfrac{\sqrt{3}}{2}$. The desired probability is $\dfrac{\frac{1}{2}}{\frac{\sqrt{3}}{2}} = \boxed{\dfrac{\sqrt{3}}{3}}$.

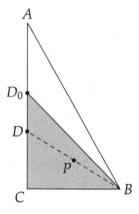

10.19

(a) Let ℓ be the line parallel to AB through the midpoint of AD. If P is between l and CD, then ABP
has a greater area than that of triangle CDP, since the height from P to AB is greater than the
height from P to CD. The region between ℓ and CD is half of $ABCD$, so the probability is $\boxed{1/2}$.

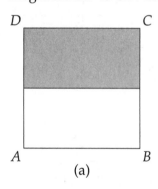 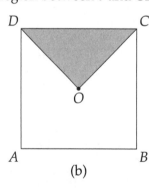 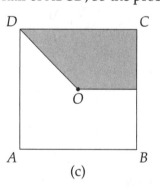

(b) Let O be the center of the square. Since ABP, BCP, CDP, and DAP all have the same base, ABP
has the greatest area if and only if P is farther from side AB than from all the other sides. In other
words, P must be in triangle OCD. The answer is $\boxed{1/4}$.

(c) As in part (a), ABP has greater area than CDP if P is closer to CD than AB, which means P must
between the side CD and the line ℓ (parallel to AB) connecting the midpoints of AD and BC.
Additionally, ABP has a greater area than BCP if P is closer to BC than AB, so P must be on the
side of diagonal BD closer to point C. The intersection of these two regions in the unit square is a
trapezoid with area $\boxed{3/8}$.

10.20 Notice that the condition will be satisfied if and only if the arcs between each pair of points are all less than or equal to 60 degrees. Let the center of the circle be O and the three points be A, B, C. Since we are looking at the relative positions of the three points, we may fix one of the points A. Then we let $x = \angle AOB$ and $y = \angle AOC$ such that $-180 < x, y \le 180$, where we think of the angle as positive if the point lies counterclockwise from A, and as negative if the point lies clockwise from A. Then for all of the arcs to be less than or equal to 60 degrees, we must have $-60 \le x \le 60$, $-60 \le y \le 60$, and $-60 \le x - y \le 60$. We graph this region in the diagram to the right, and we can see its area is $120^2 - 60^2$. The total possible area is 360^2. Thus the answer is

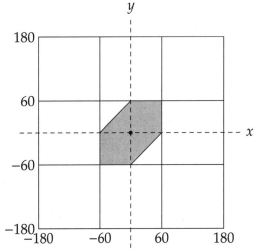

$$\frac{120^2 - 60^2}{360^2} = \frac{3}{36} = \boxed{\frac{1}{12}}.$$

10.21 Note that the triangle will contain the origin of the circle if and only if all of the arc lengths between the 3 chosen points are less than 180 degrees. Let the center of the circle be O and the three points be A, B, C. We are interested in only the relative positions of the points, so fix one of the 3 points A. If we "break" the circle apart at A as shown in the diagram below, then we can think of choosing B and C on the unit line segment, and the condition is that none of the segments AB, BC, CA are longer than $1/2$ of the entire segment AA. But this is exactly the situation that we examined in Problem 10.16 above, so the probability of success is $\boxed{1/4}$.

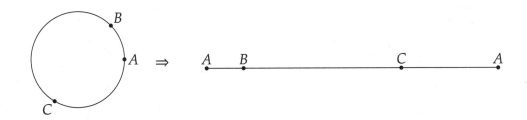

CHAPTER 11

Expected Value

Exercises for Section 11.2

11.2.1 In one flip, we have a 1/2 chance of getting heads and winning 3 dollars, and a 1/2 chance of getting tails and losing two dollars. So the expected value of one flip is $E = \frac{1}{2}(\$3) + \frac{1}{2}(-\$2) = \boxed{\$0.50}$.

11.2.2 $\boxed{\text{Yes}}$. The probability of any possible outcome is greater than zero, while the value of any outcome is also greater than zero. So the expected value of the event is the sum of

$$(\text{probability of an event}) \times (\text{value of an event})$$

summed over all events; in particular, it is a sum of positive numbers, and is thus positive.

11.2.3 The probability that value 1 occurs is p, and the probability that outcome 0 occurs is $1 - p$. So

$$E = (P(\text{event } 1) \times 1) + (P(\text{event } 0) \times 0) = (p \times 1) + ((1 - p) \times 0) = \boxed{p}.$$

Exercises for Section 11.3

11.3.1 We can make a table of the 36 equally likely possible outcomes for the sum and the product:

+	⚀	⚁	⚂	⚃	⚄	⚅
⚀	2	3	4	5	6	7
⚁	3	4	5	6	7	8
⚂	4	5	6	7	8	9
⚃	5	6	7	8	9	10
⚄	6	7	8	9	10	11
⚅	7	8	9	10	11	12

×	⚀	⚁	⚂	⚃	⚄	⚅
⚀	1	2	3	4	5	6
⚁	2	4	6	8	10	12
⚂	3	6	9	12	15	18
⚃	4	8	12	16	20	24
⚄	5	10	15	20	25	30
⚅	6	12	18	24	30	36

Since the 36 outcomes in each table are equally likely, we can average them to find the expected value. The numbers in the sum table add up to 252, and the numbers in the product table add up to 441, so the expected values are

$$E(\text{sum of two dice}) = \frac{252}{36} = \boxed{7} \quad \text{and} \quad E(\text{product of two dice}) = \frac{441}{36} = \boxed{12.25}.$$

Note the following observation: recall from the text that $E(\text{one die}) = 3.5$. Then

$$E(\text{sum of two dice}) = 7 = 3.5 + 3.5 = E(\text{one die}) + E(\text{one die}),$$

and

$$E(\text{product of two dice}) = 12.25 = 3.5 \times 3.5 = E(\text{one die}) \times E(\text{one die}).$$

Why does this work?

11.3.2 The expected value is $E = \left(\frac{1}{2} \times \$1\right) + \left(\frac{1}{3} \times \$3\right) + \left(\frac{1}{6} \times (-\$5)\right) = \$\frac{4}{6} = \boxed{\$\frac{2}{3} \approx \$0.67}$.

11.3.3 Each number from $\boxdot$ to $\boxed{\vdots\vdots}$ has probability $\frac{1}{6}$ of being rolled, so the expected value is

$$E = \left(\frac{1}{6} \times \$1^2\right) + \left(\frac{1}{6} \times \$2^2\right) + \cdots + \left(\frac{1}{6} \times \$6^2\right) = \frac{1}{6}(\$1 + \$4 + \$9 + \$16 + \$25 + \$36) = \boxed{\$\frac{91}{6} \approx \$15.17}.$$

11.3.4 Let E_1 be the expected winnings if a $\heartsuit$ or $\diamond$ is drawn. Since the probability that any particular rank is drawn is the same for any rank, the expected value is simply the average all the winnings for each rank, so

$$E_1 = \frac{1}{13}(\$1 + \$2 + \cdots + \$10 + (3 \times \$20)) = \$\frac{115}{13}.$$

Let E_2 be the expected winnings if a $\clubsuit$ is drawn and E_3 the expected winnings if a $\spadesuit$ is drawn. Since drawing a $\clubsuit$ doubles the winnings and drawing a $\spadesuit$ triples the winnings, $E_2 = 2E_1$ and $E_3 = 3E_1$. Since there is an equal chance drawing each of the suits, we can average their expected winnings to find the overall expected winnings. Therefore the expected winnings are

$$E = \frac{1}{4}(E_1 + E_1 + E_2 + E_3) = \frac{1}{4}(7E_1) = \boxed{\$\frac{805}{52}},$$

or about $\$15.48$, which is the fair price to pay to play the game.

11.3.5 There are 5 white balls and $5 + k$ total balls, so the probability that a white ball is drawn is $\frac{5}{5+k}$. Similarly, the probability that a black ball is drawn is $\frac{k}{5+k}$. So

$$E = \frac{5}{5+k}(1) + \frac{k}{5+k}(-1) = -\frac{1}{2}.$$

Multiply both sides of the equation by $2(5+k)$ to get $10 - 2k = -5 - k$, and we see that $\boxed{k = 15}$.

11.3.6 Let x be the probability that a $\boxdot$ is rolled. Then the probability that a $\boxed{\vdots}$ is rolled is $2x$, the probability that a $\boxed{\because}$ is rolled is $3x$, and so on. Since the sum of all these probabilities must be 1, we have that $x + 2x + \cdots + 6x = 1$, which means that $21x = 1$, so $x = \frac{1}{21}$. Therefore

$$E = \frac{1}{21}(1) + \frac{2}{21}(2) + \cdots + \frac{6}{21}(6) = \frac{1^2 + 2^2 + \cdots 6^2}{21} = \boxed{\frac{13}{3}}.$$

Exercises for Section 11.4

11.4.1

(a) The two envelopes could either have five and ten dollars or ten and twenty dollars with equal probability. So the other envelope has either $5 or $20 in it. Therefore, the expected value of the other envelope is $E = \frac{1}{2}(\$5 + \$20) = \boxed{\$12.50}$.

(b) Yes, the expected value of the second envelope is more than the original envelope, so we should switch.

(c) If x is in the first envelope, either $2x$ or $\frac{x}{2}$ dollars is in the second envelope. These two outcomes are equally likely, giving the other envelope an expected value of $\frac{1}{2}(\$2x + \$\frac{x}{2}) = \$\frac{5}{4}x > \$x$. So no matter how much money we discover in the first envelope, it appears that we are better off switching.

(d) Since each envelope has a 50/50 chance of being the higher amount, there clearly should be no value in switching. Or to put it another way, the strategy of always not switching should work just as well as the strategy of always switching. But our answer for part (c) seems to indicate that we should always switch! This looks like a paradox.

It actually is quite difficult to resolve the paradox; to come up with a satisfactory explanation for what's going on requires somewhat advanced probability techniques.

Review Problems

11.6 There is a $\frac{1}{2}$ probability of rolling an odd number and winning $0, and a $\frac{1}{6}$ probability of winning each of $2, $4, or $6. So $E = \frac{1}{2} \times \$0 + \frac{1}{6} \times (\$2 + \$4 + \$6) = \boxed{\$2}$.

11.7 There is a $\frac{1}{2}$ probability that each coin comes up heads, so the expected value of the coins that come up heads is $\frac{1}{2}(1¢ + 5¢ + 10¢ + 25¢) = \boxed{20.5¢}$.

11.8

(a) The expected value of one roll is the average of all the outcomes, or $E = \frac{1}{8}(1 + 2 + \cdots + 8) = \boxed{4.5}$.

(b) To find the expected value of a double roll, we can simply add the expected values of the individual rolls, giving $4.5 + 4.5 = \boxed{9}$.

11.9 Since Bin A has one white ball and four black balls, the money ball has a $\frac{1}{5}$ chance of coming from Bin W and a $\frac{4}{5}$ chance of coming from Bin B. The total expected value therefore is $E = \frac{1}{5}E_W + \frac{4}{5}E_B$, where E_W and E_B are the expected values of a ball drawn from bins W and B, respectively. Since Bin W has five $8 balls and 1 $500 ball, its expected value is

$$E_W = \frac{5}{6} \times \$8 + \frac{1}{6} \times \$500 = \$90.$$

Since Bin B has three \$1 balls and one \$7 ball, its expected value is

$$E_B = \frac{3}{4} \times \$1 + \frac{1}{4} \times \$7 = \$2.5.$$

Therefore

$$E = \frac{1}{5}E_W + \frac{4}{5}E_B = \frac{1}{5}(\$90) + \frac{4}{5}(\$2.5) = \boxed{\$20}.$$

11.10

(a) There are $\binom{5}{2} = 10$ different pairs of marbles can be drawn, and the expected value of the sum is the average of the sums of each pair. This is

$$\frac{1}{10}((1+2)+(1+3)+(1+4)+(1+5)+(2+3)+(2+4)+(2+5)+(3+4)+(3+5)+(4+5)) = \frac{60}{10} = \boxed{6}.$$

(b) There are $\binom{5}{2} = 10$ different pairs of marbles can be drawn, and the expected value of the product is the average of the products of each pair. This is

$$\frac{1}{10}((1 \times 2)+(1 \times 3)+(1 \times 4)+(1 \times 5)+(2 \times 3)+(2 \times 4)+(2 \times 5)+(3 \times 4)+(3 \times 5)+(4 \times 5)) = \frac{85}{10} = \boxed{8.5}.$$

11.11 The products of opposite faces are $1 \times 6 = 6$, $2 \times 5 = 10$, and $3 \times 4 = 12$. Therefore, after rolling the die, the outcomes $6 \times 10 = 60$, $6 \times 12 = 72$, and $10 \times 12 = 120$ are equally likely. So the expected value is just the average of these outcomes, namely $(60 + 72 + 120)/3 = \boxed{84}$.

11.12 First, note that the three darts are independent, so we need only find the expected value of one dart and triple it to get the expected total of three darts.

The circles have radii of 2, 4, and 6. The inner circle has an area of 4π. The middle band has an area of $16\pi - 4\pi = 12\pi$. The outer band has an area of $36\pi - 16\pi = 20\pi$. The total area of the dartboard is 36π. Therefore, the expected value of one dart is

$$E = \frac{4\pi}{36\pi}(20) + \frac{12\pi}{36\pi}(10) + \frac{20\pi}{36\pi}(5) = \frac{20}{9} + \frac{10}{3} + \frac{25}{9} = \frac{75}{9}.$$

Therefore the expected value of the total of three darts is $3\left(\frac{75}{9}\right) = \frac{75}{3} = \boxed{25}$.

11.13 Suppose that an event has outcomes with values $x_1, x_2, \ldots x_n$, which are all equally likely. Since there are n possible outcomes, the probability of each one occurring is $\frac{1}{n}$. Therefore, by definition, the expected value is

$$\frac{1}{n}(x_1) + \frac{1}{n}(x_2) + \cdots + \frac{1}{n}(x_n) = \frac{1}{n}(x_1 + x_2 + \cdots + x_n) = \frac{x_1 + x_2 + \cdots + x_n}{n},$$

which is just the average of the possible outcomes.

Challenge Problems

11.14

(a) Six of the eight slips have a 1 and two of the eight slips have a 3, so $E = \dfrac{6}{8} \times 1 + \dfrac{2}{8} \times 3 = \boxed{\dfrac{3}{2}}$.

(b) Now six of the nine slips have a 1 and three of the nine slips have a 3, so $E = \dfrac{6}{9} \times 1 + \dfrac{3}{9} \times 3 = \boxed{\dfrac{5}{3}}$.

(c) Now six of the ten slips have a 1 and four of the ten slips have a 3, so $E = \dfrac{6}{10} \times 1 + \dfrac{4}{10} \times 3 = \boxed{\dfrac{9}{5}}$.

(d) If we add x 3's, there will be $8 + x$ total slips in the bag. Six of the $8 + x$ slips have a 1 and $2 + x$ of the $8 + x$ slips have a 3, so

$$E = \frac{6}{8+x} \times 1 + \frac{2+x}{8+x} \times 3 = \frac{3x+12}{x+8}.$$

We want this expected value equal to 2, so we must solve $\dfrac{3x+12}{x+8} = 2$. Multiplying by $x+8$ gives us $3x + 12 = 2x + 16$, and we see that $\boxed{x = 4}$.

(e) As in part (d), we had x slips with 3's written on them, giving an expected value of $\dfrac{3x+12}{x+8}$. We want this expected value to be at least 2.5, so $\dfrac{3x+12}{x+8} \geq 2.5$. This gives us $3x + 12 \geq 2.5x + 20$, which means that $\boxed{x \geq 16}$.

11.15

(a) There are 3! different orders that $(1, 2, 3)$ can be drawn. The chance that all 3 match the order in which they are drawn is $\dfrac{1}{3!} = \dfrac{1}{6}$. There is no way to select exactly two in the appropriate order, because then the remaining number has to be in its matching spot as well. There are three different ways to have exactly one in the appropriate order: $((1, 3, 2), (3, 2, 1), (2, 1, 3))$, so this has a $\dfrac{3}{6} = \dfrac{1}{2}$ probability. There are two different ways to have no integers in the correct order: $((2, 3, 1), (3, 1, 2))$, for a probability of $\dfrac{1}{3}$. So $E = \dfrac{1}{6} \times \$3 + \dfrac{1}{2} \times \$1 + \dfrac{1}{3} \times \$0 = \boxed{\$1}$.

(b) There are 4! different orders that $(1, 2, 3, 4)$ can be drawn.

Case 1: All 4 are in the correct place. There is one way to do this, so the probability of this happening is $\dfrac{1}{4!} = \dfrac{1}{24}$.

Case 2: Exactly 3 of the 4 are in the correct place. There is no way to select only three in the appropriate order, because the only place the remaining number has to go is its matching spot.

Case 3: Exactly 2 of the 4 are in the correct place. There are $\binom{4}{2} = 6$ ways to select exactly 2 in the appropriate order: $((1, 2, 4, 3), (1, 4, 3, 2), (1, 3, 2, 4), (4, 2, 3, 1), (3, 2, 1, 4), (2, 1, 3, 4))$, one for each pair of numbers. The probability of this case is $\dfrac{6}{24} = \dfrac{1}{4}$.

Case 4: Exactly 1 of the 4 is in the correct place. There are 4 different numbers that could be in the correct place, and for each of these, there are two different ways to order the remaining 3 numbers so that none of the other 3 are in the correct place (for example, $(1, 3, 4, 2)$ and $(1, 4, 2, 3)$). So there are $4 \times 2 = 8$ ways to do this, and the probability of it happening is $\frac{8}{24} = \frac{1}{3}$.

Case 5: None of the numbers is in the correct place. There are $24 - 1 - 6 - 8 = 9$ remaining ways do this, so the probability is $\frac{9}{24} = \frac{3}{8}$.

So $E = \frac{1}{24} \times \$4 + \frac{1}{4} \times \$2 + \frac{1}{3} \times \$1 + \frac{3}{8} \times \$0 = \boxed{\$1}$.

(c) The answer should be $1 for any number of balls n. Looking at any individual draw, there are n equally likely possibilities for which ball is drawn, so the probability that the correct one is chosen is $\frac{1}{n}$. Since n balls are chosen in each sequence of draws, we have $E = n \times \left(\frac{1}{n} \times \$1\right) = \boxed{\$1}$.

11.16 The area of any such triangle chosen is $\frac{1}{2}xy$. There are 25 different ordered pairs of x and y that can be chosen: $((1, 1), (1, 2), (1, 3), \ldots, (5, 5))$. Each of these is chosen with equal probability, so

$$E = \frac{1}{25}\left(\frac{1}{2}(1 \times 1) + \frac{1}{2}(1 \times 2) + \frac{1}{2}(1 \times 3) + \cdots + \frac{1}{2}(5 \times 5)\right)$$
$$= \frac{1}{50}(1(1 + 2 + 3 + 4 + 5) + 2(1 + 2 + 3 + 4 + 5) + \cdots + 5(1 + 2 + 3 + 4 + 5))$$
$$= \frac{1}{50}((1 + 2 + 3 + 4 + 5)(1 + 2 + 3 + 4 + 5))$$
$$= \frac{1}{50}(15^2) = \boxed{\frac{9}{2}}.$$

11.17 The best strategy is to stop drawing balls if you draw a white ball first, and if a black ball is drawn first, continue drawing until both white balls are removed.

We will examine the expected outcome of this strategy. Note that any sequence has a $\frac{1}{\binom{5}{2}} = \frac{1}{10}$ probability of occurring. To the right, we list the 10 possible sequences of draws; parentheses will be placed after the strategy tells the player to quit.

Drawing	Win	Drawing	Win
W(WBBB)	$1	W(BWBB)	$1
W(BBWB)	$1	W(BBBW)	$1
BWW(BB)	$1	BWBW(B)	$0
BWBBW	-$1	BBWW(B)	$0
BBWBW	-$1	BBBWW	-$1

In $\frac{5}{10} = \frac{1}{2}$ of the sequences, you win 1 dollar; in $\frac{2}{10} = \frac{1}{5}$ of the sequences, you break even; and in $\frac{3}{10}$ of the sequences, you lose a dollar. So

$$E = \frac{1}{2} \times \$1 + \frac{1}{5} \times \$0 + \frac{3}{10} \times (-\$1) = \$\frac{1}{5}.$$

11.18 The probability that the coin first comes up heads on the first flip is p. The probability that the coin first comes up heads on the second flip is $(1 - p)p$, because it must come up tails on the first flip. More generally, the probability that the coin first comes up heads after n flips is $(1 - p)^{n-1}p$, because we must flip $n - 1$ tails followed by a head. So multiplying each probability by the number of flips, we find the expected value is

$$E = (1 \times p) + (2 \times p(1 - p)) + (3 \times p(1 - p)^2) + \cdots + ((n + 1) \times p(1 - p)^n) + \cdots.$$

Now we must simplify this series. Note that

$$(1 - p)E = (1 \times p(1 - p)) + (2 \times p(1 - p)^2) + (3 \times p(1 - p)^3) + \cdots + ((n + 1) \times p(1 - p)^{n+1}) + \cdots.$$

Subtracting these two equations, we get that

$$E - (1 - p)E = pE = p + p(1 - p) + p(1 - p)^2 + \cdots + p(1 - p)^n + \cdots.$$

Dividing by p, we get

$$E = 1 + (1 - p) + (1 - p)^2 + \cdots + (1 - p)^n + \cdots.$$

This is a geometric series with initial term 1 and ratio $1 - p$, so $E = \dfrac{1}{1 - (1 - p)} = \boxed{\dfrac{1}{p}}$.

11.19   If we hire 1 helicopter, we have a 0.9 probability of recovering the treasure. So our expected gain from finding the treasure is $(0.9) \times \$100{,}000 = \$90{,}000$. After subtracting the \$1,000 cost of hiring the helicopter, we are left with a net expected gain of \$89,000.

If we hire 2 helicopters, then the probability of recovering the treasure is the complement of the probability of both helicopters failing to find it, which gives $1 - (0.1)^2 = 0.99$. After subtracting the cost of the helicopters, this brings a net expected gain of $(0.99) \times (\$100{,}000) - \$2{,}000 = \$97{,}000$.

If we hire 3 or more helicopters, then the cost of the helicopters is at least \$3,000, and since there is never any guarantee that the treasure is recovered, the net expected value of return of the treasure must be less than $\$100{,}000 - \$3{,}000 = \$97{,}000$.

So $\boxed{2}$ is the ideal number of helicopters to hire.

11.20   Let the probability that Caitlin wins be p. Eventually, one of them has to win, so the probability that Olivia wins is $1 - p$. If Caitlin wins, she ends up with 1500 coins and if she loses she ends up with 0 coins, so her expected number of coins at the end of the game is $E = p \times 1500 + (1 - p) \times 0 = 1500p$. But since on every turn, Caitlin has an equal chance of gaining or losing a penny, the expected value of the number of coins that Caitlin has never changes, and stays at 1000 coins (the number of coins that she starts with). So $E = 1000 = 1500p$, which means $p = \boxed{\dfrac{2}{3}}$.

12

Pascal's Triangle

Exercises for Section 12.3

12.3.1 The number of paths to C is given by $\binom{4}{1} = 4$, because we need to take 4 steps, 3 to the left and 1 to the right. Similarly, the number of paths to D is $\binom{4}{2} = 6$, because we need to take 2 steps left and 2 steps right. The number of paths to E is $\binom{5}{2} = 10$, because we need to take 3 steps left and 2 steps right. Indeed $4 + 6 = 10$.

12.3.2 The number of paths to F is $\binom{5}{3} = 10$, because we must take 2 steps left and 3 steps right. Similarly, the number of paths to G is $\binom{5}{4} = 5$ (1 step left and 4 steps right), and the number of paths to H is $\binom{6}{4} = 15$ (2 steps left and 4 steps right). Indeed $10 + 5 = 15$.

12.3.3 There is only one path to the left edge of the row (all steps must be to the left), so the entry there is given by $\binom{n}{0} = 1$. Similarly, there are n paths to the next position in the row (we must choose 1 of our n steps to be to the right), so the entry there is $\binom{n}{1} = n$.

Exercises for Section 12.4

12.4.1

(a) Proof using algebra:

$$\binom{n}{m}\binom{m}{r} = \frac{n!}{\cancel{m!}(n-m)!} \times \frac{\cancel{m!}}{r!(m-r)!} = \frac{n!}{(n-m)!r!(m-r)!}$$

$$\binom{n}{r}\binom{n-r}{m-r} = \frac{n!}{r!\cancel{(n-r)!}} \times \frac{\cancel{(n-r)!}}{(m-r)!(n-m)!} = \frac{n!}{r!(m-r)!(n-m)!}.$$

Proof using committee-forming: We wish to choose an m-person committee from n people, and then choose an r-person subcommittee from this m-person committee.

There are $\binom{n}{m}$ ways to choose from n people an m-person committee, and $\binom{m}{r}$ ways to choose from this m-person committee an r-person subcommittee. Thus the the total number of ways is $\binom{n}{m}\binom{m}{r}$.

On the other hand, we can accomplish the same task in a different way. We can first choose the r-person subcommittee from n people, and then choose from the $(n-r)$ remaining people the other $(m - r)$ members of the m-person committee. The total number of choices using this second method is $\binom{n}{r}\binom{n-r}{m-r}$.

(b) Proof using committee-forming: We wish to choose an r-person committee from m men and n women. The number of ways to choose an r-person committee from $(m + n)$ people is $\binom{m+n}{r}$.

On the other hand, the committee can have k men and $(r - k)$ women for any k from 0 to r inclusive. There are $\binom{m}{k}$ ways to choose k men (from the m total men) and $\binom{n}{r-k}$ ways to choose $(r - k)$ women (from the n total women). The total number of committees given by this second method of counting is

$$\binom{m}{0}\binom{n}{r} + \binom{m}{1}\binom{n}{r-1} + \binom{m}{2}\binom{n}{r-2} + \cdots + \binom{m}{r}\binom{n}{0}.$$

Proof using block-walking: We are trying to count the number of ways to move r steps to the right out of $m + n$ steps, or $\binom{m+n}{r}$. After m steps down the Pascals Triangle, we could have taken 0, 1, 2, ..., or r steps to the right. If we take k steps to the right after m steps, we must take $r - k$ steps to the right in the final n steps. There are $\binom{m}{k} \times \binom{n}{r-k}$ ways to do this. Summing this over k from 0 to r, we get

$$\binom{m}{0}\binom{n}{r} + \binom{m}{1}\binom{n}{r-1} + \binom{m}{2}\binom{n}{r-2} + \cdots + \binom{m}{r}\binom{n}{0}.$$

(c) Proof using block-walking: The right side $\binom{2n}{n}$ counts the number of paths from the top of Pascal's Triangle to the middle point of Row $2n$. Let us count the number of such paths which pass through the point which is k spots from the left edge of Row n. There are $\binom{n}{k}$ paths from the top of Pascal's Triangle to point k of Row n, since we have to take n total steps, k of which are to the right. Then there are $\binom{n}{k}$ paths from this point to the middle point of Row $2n$, since we have to take n total steps, k of which are to the left. So there are $\binom{n}{k} \times \binom{n}{k} = \binom{n}{k}^2$ paths from the top of Pascal's Triangle to the middle point of Row $2n$ which pass through point k of Row n. But every path to the middle point of Row $2n$ must pass through exactly one of the points on Row n, so we can add the number of such paths and get

$$\binom{2n}{n} = \binom{n}{0}^2 + \binom{n}{1}^2 + \cdots + \binom{n}{n}^2,$$

as desired.

Proof using algebra: Recall that $\binom{n}{k} = \binom{n}{n-k}$. Thus $\binom{n}{k}\binom{n}{n-k} = \binom{n}{k}^2$. In the equation from part (b), set $m = n = r$. Then the equation becomes

$$\binom{n}{0}\binom{n}{n} + \binom{n}{1}\binom{n}{n-1} + \binom{n}{2}\binom{n}{n-2} + \cdots + \binom{n}{n}\binom{n}{0} = \binom{2n}{n}.$$

Thus

$$\binom{n}{0}^2 + \binom{n}{1}^2 + \binom{n}{2}^2 + \cdots + \binom{n}{n}^2 = \binom{2n}{n}.$$

(d) We can use Pascal's Identity on the left side to get

$$\binom{2n}{n} + \binom{2n}{n-1} = \binom{2n+1}{n}.$$

But note that $\binom{2n+1}{n} = \binom{2n+1}{(2n+1)-n} = \binom{2n+1}{n+1}$. Therefore,

$$\binom{2n+1}{n} = \frac{1}{2}\left(\binom{2n+1}{n} + \binom{2n+1}{n+1}\right),$$

and we can apply Pascal's Identity again to get

$$\binom{2n+1}{n} = \frac{1}{2}\binom{2n+2}{n+1},$$

completing the proof.

Exercises for Section 12.5

12.5.1 We wish to choose from n people a committee, but we don't care what size the committee is (it could even be empty). Since each person has 2 choices—they can either be on or not on the committee—the number of ways to form a committee is 2^n.

On the other hand, the committee can have k people for any k from 0 to n inclusive. There are $\binom{n}{k}$ ways to choose from n people a committee of k people. The total number of committees given by this second way of counting is

$$\binom{n}{0} + \binom{n}{1} + \binom{n}{2} + \cdots + \binom{n}{n}.$$

Therefore,

$$\binom{n}{0} + \binom{n}{1} + \binom{n}{2} + \cdots + \binom{n}{n} = 2^n.$$

12.5.2 We already know that

$$\binom{n}{0} + \binom{n}{1} + \binom{n}{2} + \cdots + \binom{n}{n} = 2^n,$$

so squaring both sides gives

$$\left(\binom{n}{0} + \binom{n}{1} + \binom{n}{2} + \cdots + \binom{n}{n}\right)^2 = (2^n)^2 = 2^{2n}.$$

But on the other hand, by the reasoning in the previous problem,

$$\binom{2n}{0} + \binom{2n}{1} + \binom{2n}{2} + \cdots + \binom{2n}{2n} = 2^{2n},$$

so we're done.

12.5.3

(a) We want to show that

$$\binom{n}{0} + \binom{n}{2} + \binom{n}{4} + \cdots = \binom{n}{1} + \binom{n}{3} + \binom{n}{5} + \cdots .$$

In other words, we want to show that in the nth row, the sum of the numbers of paths to the even entries (that is, $\binom{n}{k}$ such that k is even) is equal to the sum of the numbers of paths to the odd entries. But for every path to a point in Row $(n-1)$ of Pascal's triangle, in their last step to Row n, half of them go to an even entry in Row n and half of them go to an odd entry in Row n. Thus the total number of paths to all even entries must equal the total number of paths all odd entries.

(b) There are n people, and a committee can have any number of people from 0 to n inclusive. We want to show the number of committees having an odd number of people is equal to the number of committees having an even number of people. Let A be one of the n people. We will pair up all of the even and odd committees, which will prove the desired identity.

If a committee C has person A, pair it up with the committee that is $C - A$ (person A removed from C). Equivalently, if a committee C does not have person A, pair it up with the committee that is $C \cup A$ (person A is put on committee C). Since such a pairing always creates two committees that differ by 1 in the number of people, we can see one of the two is even and the other is odd.

Therefore, since the even committees and the odd committees can be paired, each makes up half of the number of total committees. In particular, the number of even committees is equal to the number of odd committees.

Review Problems

12.11

(a) By swapping left and right moves, any path to the rth entry of Row n can be turned into a path to the $(n - r)$th entry of Row n. Conversely, this same swapping also turns any path to the $(n - r)$th entry of Row n into a path to the rth entry of Row n. Thus the number of paths to the rth entry of Row n is equal to the the number of paths to the $(n - r)$th entry of the Row n, or in other words, $\binom{n}{r} = \binom{n}{n-r}$.

(b) The number of ways to choose an r-person committee from n people is $\binom{n}{r}$. But we could instead choose the $n - r$ people who are *not* on an r-person committee; the number of ways to do this is $\binom{n}{n-r}$. Both count the same thing, so we must have $\binom{n}{r} = \binom{n}{n-r}$

(c)

$$\binom{n}{r} = \frac{n!}{r!(n-r)!} = \frac{n!}{(n-r)!r!} = \frac{n!}{(n-r)!(n-(n-r))!} = \binom{n}{n-r}.$$

12.12

(a)

$$\frac{r}{n}\binom{n}{r} = \frac{r}{n} \times \frac{n!}{r!(n-r)!} = \frac{(n-1)!}{(r-1)!(n-r)!} = \binom{n-1}{r-1}.$$

(b) We wish to choose an r-person committee from n people, then choose a president from the r people on the committee. We can do this in $r\binom{n}{r}$ ways. On the other hand, we can first pick the president from the original n people, and then pick the other $(r-1)$ committee members of the committee from the $n-1$ remaining people. We can do this in $n\binom{n-1}{r-1}$ ways. Both ways count the same thing, so $r\binom{n}{r} = n\binom{n-1}{r-1}$, and hence $\binom{n-1}{r-1} = \frac{r}{n}\binom{n}{r}$.

12.13

(a) By swapping steps to the left and steps to the right, any path to the nth, or middle, entry of Row $2n$ can be turned into a different path to the nth entry of Row $2n$. Thus the number of paths to the nth entry of Row $2n$ come in pairs, and hence $\binom{2n}{n}$ is even.

(b) Consider all n-person committees formed from $2n$ people. We can pair these committees by matching two committees A and B if A and B have no common member (in other words, A and B combined comprises all $2n$ people). Therefore the number of n-person committees, namely $\binom{2n}{n}$, is even.

(c) By the identity from Problem 12.12:

$$\binom{2n}{n} = \frac{2n}{n}\binom{2n-1}{n-1} = 2\binom{2n-1}{n-1}.$$

Since $\binom{2n}{n}$ is 2 times an integer, it must be even.

12.14

(a)

$$2\binom{n}{2} + n^2 = n(n-1) + n^2 = 2n^2 - n = \frac{(2n)(2n-1)}{2} = \binom{2n}{2}.$$

(b) Consider choosing a 2-person committee from n boys and n girls. Since there are $2n$ total people, the number of ways to form this committee is $\binom{2n}{2}$. On the other hand, we can break up the possible committees into cases.

Case 1: Committees with 2 boys. There are $\binom{n}{2}$ of these, since we must choose 2 out of n boys.

Case 2: Committees with 2 girls. There are $\binom{n}{2}$ of these, since we must choose 2 out of n girls.

Case 3: Committees with 1 boy and 1 girl. There are n choices for the boy and n choices for the girl, so there are n^2 of these committees.

We add our cases, and thus $\binom{2n}{2} = 2\binom{n}{2} + n^2$.

12.15

(a) Proof by algebra:

$$\binom{n}{r} = \binom{n}{n-r} = \frac{n}{n-r}\binom{n-1}{n-r-1} = \frac{n}{n-r}\binom{n-1}{r}.$$

Proof by committee-forming: We count the ways to select a committee of r people and a president (not on the committee) from a club of n people. One way do to this is choose the r-person committee first (in $\binom{n}{r}$ ways) and then choose the president from the remaining $n-r$ people, which gives a total of $(n-r)\binom{n}{r}$ possibilities. On the other hand, we could select the president first out of the n people, and then select the r-person committee from the $n-1$ remaining people, for a total of $n\binom{n-1}{r}$ possibilities. So $(n-r)\binom{n}{r} = n\binom{n-1}{r}$, which means $\binom{n}{r} = \frac{n}{n-r}\binom{n-1}{r}$.

(b) Not an identity. For example, set $n = r = 2$, then we have

$$\binom{2}{0} + \binom{2}{1} + \binom{2}{2} = 4 \qquad \text{whereas} \qquad \binom{4}{2} = 6.$$

(c) Not an identity. For example, set $n = 2, r = m = 1, s = 0$. Then we have

$$\binom{2}{1}\binom{1}{0} = 2 \qquad \text{whereas} \qquad \binom{2}{0} = 1.$$

Challenge Problems

12.16

(a) If $\binom{n}{k}$ is odd, then $\binom{n}{n-k}$ is odd too, since Pascal's Identity states that $\binom{n}{k} = \binom{n}{n-k}$. If $k \neq n - k$, then these are different entries of Row n of Pascal's Triangle but if $k = n - k$, then $2k = n$, and we know that $\binom{2k}{k}$ is always even, as shown in Problem 12.13. Therefore the odd entries of Row n come in pairs, so the number of odd entries is even.

(b) Recall from previous exercises that

$$\binom{n}{0} + \binom{n}{1} + \binom{n}{2} + \cdots + \binom{n}{n} = 2^n$$

and

$$\binom{n}{0} + \binom{n}{2} + \binom{n}{4} + \cdots = \binom{n}{1} + \binom{n}{3} + \binom{n}{5} + \cdots.$$

Therefore

$$
\begin{aligned}
2^n &= \binom{n}{0} + \binom{n}{1} + \binom{n}{2} + \cdots + \binom{n}{n} \\
&= \left(\binom{n}{0} + \binom{n}{2} + \binom{n}{4} + \cdots \right) + \left(\binom{n}{1} + \binom{n}{3} + \binom{n}{5} + \cdots \right) \\
&= 2\left(\binom{n}{0} + \binom{n}{2} + \binom{n}{4} + \cdots \right).
\end{aligned}
$$

So

$$\binom{n}{0} + \binom{n}{2} + \binom{n}{4} + \cdots = \frac{1}{2}(2^n) = \boxed{2^{n-1}}.$$

12.17 We can rewrite the sum as

$$\binom{n}{0}\binom{n}{n-1} + \binom{n}{1}\binom{n}{n-2} + \cdots + \binom{n}{n-1}\binom{n}{0}.$$

The general term is of the form $\binom{n}{k}\binom{n}{n-1-k}$. We can use a committee-forming argument: think of choosing of committee of size $n - 1$ from a group of $2n$ people, n of whom are men and n of whom are women. There are $\binom{2n}{n-1}$ ways to do this. But if the committee has k men and $n - 1 - k$ women, there are $\binom{n}{k}\binom{n}{n-1-k}$

ways to choose it. So the left side of the desired equation sums the number of ways to choose committees by gender makeup, and the right side just counts the number of committees without regard to gender.

12.18 Consider the ratio of 3 consecutive entries $\binom{n}{k-1}$, $\binom{n}{k}$, $\binom{n}{k+1}$, which we can write out as

$$\frac{n!}{(k-1)!(n-k+1)!} : \frac{n!}{k!(n-k)!} : \frac{n!}{(k+1)!(n-k-1)!}.$$

If we cancel $n!$, and multiply each term by $(k+1)!(n-k+1)!$, we're left with the ratio

$$k(k+1) : (k+1)(n-k+1) : (n-k)(n-k+1).$$

Since this ratio must equal $3 : 4 : 5$, we want to solve the system

$$4k(k+1) = 3(k+1)(n-k+1)$$
$$5(k+1)(n-k+1) = 4(n-k)(n-k+1).$$

The system simplifies to

$$4k = 3(n-k+1)$$
$$5(k+1) = 4(n-k).$$

This easily solves to give $n = \boxed{62}$, $k = 27$.

12.19 The six numbers surrounding the kth entry on row n are

$$\binom{n-1}{k-1}, \binom{n-1}{k}, \binom{n}{k+1}, \binom{n+1}{k+1}, \binom{n+1}{k}, \binom{n}{k-1}.$$

The pattern is not a coincidence:

$$\binom{n-1}{k-1}\binom{n}{k+1}\binom{n+1}{k} = \frac{(n-1)!}{(k-1)!(n-k)!} \times \frac{n!}{(k+1)!(n-k-1)!} \times \frac{(n+1)!}{k!(n-k+1)!},$$

whereas

$$\binom{n-1}{k}\binom{n+1}{k+1}\binom{n}{k-1} = \frac{(n-1)!}{k!(n-k-1)!} \times \frac{(n+1)!}{(k+1)!(n-k)!} \times \frac{n!}{(k-1)!(n-k+1)!}.$$

These products have the same terms in the numerators and denominators, thus the two products are equal.

12.20 The Fibonacci numbers are given by $F_0 = 1$, $F_1 = 1$, and $F_n = F_{n-1} + F_{n-2}$. After some exploration, we find that

$$F_n = \binom{n}{0} + \binom{n-1}{1} + \binom{n-2}{2} + \cdots.$$

This sum is a skewed diagonal in Pascal's triangle, as shown in the following picture:

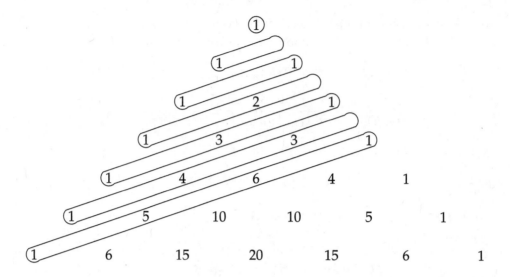

Let's verify $F_n + F_{n+1} = F_{n+2}$.

$$F_n + F_{n+1} = \left(\binom{n}{0} + \binom{n-1}{1} + \binom{n-2}{2} + \cdots\right) + \left(\binom{n+1}{0} + \binom{n}{1} + \binom{n-1}{2} + \cdots\right)$$

$$= \binom{n+1}{0} + \left(\binom{n}{0} + \binom{n}{1}\right) + \left(\binom{n-1}{1} + \binom{n-1}{2}\right) + \cdots$$

$$= \binom{n+2}{0} + \binom{n+1}{1} + \binom{n}{2} + \cdots$$

$$= F_{n+2}.$$

CHAPTER 13

The Hockey Stick Identity

Exercises for Section 13.3

13.3.1 The number of ways to distribute 10 pieces of candy to 5 kids is

$$\binom{10+5-1}{5-1} = \boxed{1001}.$$

13.3.2 We first give 1 piece of candy to each kid, to make sure that each kid gets a piece, then we distribute the remaining 7 pieces of candy to the 4 kids. The answer is

$$\binom{7+4-1}{4-1} = \boxed{120}.$$

13.3.3 Distributions of the two kinds of candy are independent. Thus the answer is

$$\binom{5+4-1}{4-1}\binom{6+4-1}{4-1} = \boxed{4704}.$$

Exercises for Section 13.4

13.4.1 The number of ways to distribute 16 pieces of candy to 6 kids is

$$\binom{16+6-1}{6-1} = \boxed{20{,}349}.$$

13.4.2 We first give 2 pieces of candy to the oldest kid, then distribute the remaining 11 pieces of candy to the 5 kids. The answer is

$$\binom{11+5-1}{5-1} = \binom{15}{4} = \boxed{1365}.$$

13.4.3 If the twins each get k pieces, we must distribute the remaining $(12-2k)$ pieces of candy to the other two kids. There are $(12-2k)+1 = 13-2k$ ways that we can do this: the first remaining kid can get any amount of $0, 1, 2, \ldots, 12-2k$ pieces, and the other kid gets the rest. Since k can be any integer from 0 to 6 inclusive, the answer is $13 + 11 + 9 + 7 + 5 + 3 + 1 = \boxed{49}$.

Exercises for Section 13.5

13.5.1

(a) This is a straightforward application of the Hockey Stick Identity:

$$\binom{2}{2} + \binom{3}{2} + \cdots + \binom{7}{2} = \binom{8}{3} = \boxed{56}.$$

(b) We can't apply the identity directly, since we're missing the $\binom{3}{3}$ and $\binom{4}{3}$ terms. But we can add those terms back in:

$$\binom{5}{3} + \binom{6}{3} + \cdots + \binom{10}{3} = \left(\binom{3}{3} + \binom{4}{3} + \cdots + \binom{10}{3}\right) - \left(\binom{3}{3} + \binom{4}{3}\right)$$

$$= \binom{11}{4} - \binom{5}{4} = \boxed{325}.$$

(c) Here we have to use the identity $\binom{n}{k} = \binom{n}{n-k}$ first before we can apply the Hockey Stick Identity:

$$\binom{6}{0} + \binom{7}{1} + \binom{8}{2} + \cdots + \binom{15}{9} = \binom{6}{6} + \binom{7}{6} + \binom{8}{6} + \cdots + \binom{15}{6} = \binom{16}{7} = \boxed{11{,}440}.$$

13.5.2

(a) $\binom{n+1}{r+1}$ counts the number of ways to move $r + 1$ steps to the right in a total of $n + 1$ steps. Each path takes its last step to the right in some row between $r + 1$ and $n + 1$. If it takes its last step to the right at row k, then there are $\binom{k-1}{r}$ such paths, since we are arranging r rights among the first $k - 1$ steps; the rest of the steps in the path will all be lefts, which can only be done in 1 way. Summing k from $r + 1$ to $n + 1$, we get $\binom{n+1}{r+1} = \binom{r}{r} + \binom{r+1}{r} + \cdots + \binom{n}{r}$.

(b) We know that $\binom{n+1}{r+1}$ counts the number of ways to choose a committee of $r + 1$ people from a total of $n + 1$ people. Rank the $n + 1$ people from 1 to $n + 1$. Each of the $r + 1$-person committees has a highest ranking member, which is some integer between $r + 1$ and $n + 1$. If the highest ranking member is k, there are $\binom{k-1}{r}$ ways to form the rest of the committee, because only people of rank $k - 1$ or less can be one of the remaining r members. Summing $\binom{k-1}{r}$ from $k = r + 1$ to $k = n + 1$ gives $\binom{n+1}{r+1} = \binom{r}{r} + \binom{r+1}{r} + \cdots + \binom{n}{r}$.

Review Problems

13.4

(a) $\dbinom{11 + 4 - 1}{4 - 1} = \boxed{364}$.

(b) $\dbinom{18 + 7 - 1}{7 - 1} = \boxed{134{,}596}$.

(c) Each kid must receive one piece of candy, so we initially distribute these first 6 pieces. Now there are 16 pieces left over, and we may distribute them to the 6 kids without restriction. Therefore the answer is $\binom{16+6-1}{6-1} = \boxed{20{,}349}$.

(d) $\binom{8+10-1}{10-1} = \boxed{24{,}310}$. Note that there are fewer pieces of candy than there are kids, but this doesn't alter the calculation.

(e) The youngest kid can have k pieces of candy, where k can be 0, 1, 2, or 3. For each of these four cases, we can first give k pieces of candy to all 5 kids, then distribute the remaining $(16-5k)$ pieces of candy to the 4 older kids in $\binom{16-5k+4-1}{4-1} = \binom{19-5k}{3}$ ways. Thus the answer is

$$\binom{19}{3} + \binom{14}{3} + \binom{9}{3} + \binom{4}{3} = \boxed{1421}.$$

13.5

(a) This is just the Hockey Stick Identity, so the answer is $\binom{10}{5} = \boxed{252}$.

(b) $\binom{6}{3} + \binom{7}{3} + \binom{8}{3} + \cdots + \binom{11}{3} = \left(\binom{3}{3} + \binom{4}{3} + \binom{5}{3} + \binom{6}{3} + \binom{7}{3} + \binom{8}{3} + \cdots + \binom{11}{3}\right) - \left(\binom{3}{3} + \binom{4}{3} + \binom{5}{3}\right)$. Applying the Hockey Stick Identity twice, we get $\binom{12}{4} - \binom{6}{4} = \boxed{480}$.

(c) All the terms in the sum are in the form $\binom{n+3}{n}$, which is equal to $\binom{n+3}{3}$. So we can convert this sum to the form $\binom{3}{3} + \binom{4}{3} + \binom{5}{3} + \cdots + \binom{10}{3}$, which by the Hockey Stick Identity is $\binom{11}{4} = \boxed{330}$.

(d) This is simply the Hockey Stick Identity, giving $\boxed{\binom{2n+1}{n+1}}$.

13.6 First, give 2 of each kind of donuts to each person, so we have 8 chocolate, 5 glazed, and 7 jelly left to distribute. Each distribution is independent of the other two. The chocolates can be distributed in $\binom{8+4-1}{4-1} = \binom{11}{3} = 165$ ways. The glazed can be distributed in $\binom{5+4-1}{4-1} = \binom{8}{3} = 56$ ways. The jellies can be distributed in $\binom{7+4-1}{4-1} = \binom{10}{3} = 120$ ways. Therefore there are a total of $165 \times 56 \times 120 = \boxed{1{,}108{,}800}$ ways to distribute the donuts.

13.7

(a) We can think of this as distributing 70 votes among the 3 candidates. This can be done in $\binom{70+3-1}{3-1} = \binom{72}{2} = \boxed{2556}$ ways.

(b) Now we add a fourth "candidate" to represent the students who do not vote. So now the votes can be distributed in $\binom{70+4-1}{4-1} = \binom{73}{3} = \boxed{62{,}196}$ ways.

Challenge Problems

13.8

(a) Not an identity. Set $r = 0$ and $n = 2$. Then the "identity" claims that $\binom{0}{0} + \binom{1}{2} + \binom{2}{4} = \binom{3}{4}$, but $1 + 0 + 0 \neq 0$.

(b) This is an identity. Note that all of the terms on the left side are of the form $k(k-1)(k-2)$, which is equal to $6\binom{k}{3}$. So we have

$$6\binom{3}{3} + 6\binom{4}{3} + \cdots + 6\binom{n}{3} = 6\left(\binom{3}{3} + \binom{4}{3} + \cdots + \binom{n}{3}\right) = 6\binom{n+1}{4},$$

where the last step follows from the Hockey Stick Identity.

(c) Not an identity. Set $n = 1$. Then the "identity" claims that $\binom{1}{0}^3 + \binom{1}{1}^3 = \binom{1}{1}$, but $1 + 1 \neq 1$.

13.9

(a) This problem is equivalent to giving n pieces of candy to 3 kids. The answer is $\binom{n+3-1}{3-1} = \boxed{\binom{n+2}{2}}$.

(b) This problem is equivalent to giving n pieces of candy to 4 kids. The answer is $\binom{n+4-1}{4-1} = \boxed{\binom{n+3}{3}}$.

(c) The number of k nonnegative integers that sum to n is equivalent to the number of ways to give n pieces of candy to k kids. The answer is $\boxed{\binom{n+k-1}{k-1}}$.

13.10   We can rephrase this problem as counting the number of ways to distribute 7 pieces of taffy to 5 kids, such than none gets more than 3. Then we give each kid enough licorice so that each kid has exactly 3 pieces.

If we didn't have the "no more than 3 pieces" restriction, then the number of ways to distribute 7 pieces of taffy to 5 kids is $\binom{7+5-1}{5-1} = \binom{11}{4} = 330$. But we need to exclude the cases where some kid gets 4 or more pieces. Note that only one kid could receive more than 3 pieces, and we have 5 choices of which kid that is. Once we allocate 4 pieces to that kid, we have to distribute the remaining 3 pieces among the 5 kids, which we can do in $\binom{3+5-1}{5-1} = \binom{7}{4} = 35$ ways.

Thus there are $330 - (5 \times 35) = 155$ ways to distribute the taffy. Once we distribute the taffy, the licorice distribution is forced: we give each kid enough licorice so that he or she has 3 pieces total. So there are $\boxed{155}$ ways to distribute the candy.

13.11 We can break apart this sum into several separate Hockey Sticks:

$$12\binom{3}{3} + 11\binom{4}{3} + 10\binom{5}{3} + \cdots + 2\binom{13}{3} + \binom{14}{3} = \left(\binom{3}{3} + \binom{4}{3} + \binom{5}{3} + \cdots + \binom{13}{3} + \binom{14}{3}\right)$$
$$+ \left(\binom{3}{3} + \binom{4}{3} + \binom{5}{3} + \cdots + \binom{12}{3} + \binom{13}{3}\right)$$
$$+ \left(\binom{3}{3} + \binom{4}{3} + \binom{5}{3} + \cdots + \binom{11}{3} + \binom{12}{3}\right)$$
$$+ \quad \vdots$$
$$+ \left(\binom{3}{3} + \binom{4}{3}\right)$$
$$+ \left(\binom{3}{3}\right).$$

Now we can apply the Hockey Stick Identity multiple times on the right sides of the above expression, to get

$$\binom{15}{4} + \binom{14}{4} + \binom{13}{4} + \cdots + \binom{4}{4}.$$

Applying the Hockey Stick Identity one final time to the last expression, we get $\boxed{\binom{16}{5} = 4368}$.

13.12 We will do this by finding k^2 in terms of combinations. Note that $\binom{k}{2} = \frac{k(k-1)}{2}$ gives

$$k^2 = 2\left(\frac{k(k-1)}{2}\right) + k = 2\binom{k}{2} + \binom{k}{1}.$$

So

$$1^2 + 2^2 + \cdots + n^2 = 2\binom{1}{2} + \binom{1}{1} + 2\binom{2}{2} + \binom{2}{1} + \cdots + 2\binom{n}{2} + \binom{n}{1}$$
$$= 2\left(\binom{1}{2} + \binom{2}{2} + \cdots + \binom{n}{2}\right) + \left(\binom{1}{1} + \binom{2}{1} + \cdots + \binom{n}{1}\right).$$

Applying the hockey stick identity to both terms in the parentheses, we get

$$2\binom{n+1}{3} + \binom{n+1}{2} = \frac{(n+1)(n)(n-1)}{3} + \frac{(n+1)(n)}{2} = \boxed{\frac{n(n+1)(2n+1)}{6}}.$$

13.13 This expression has several terms of a Hockey Stick, so we will add and subtract the ones that are missing:

$$\binom{m}{r} + \binom{m+1}{r} + \cdots + \binom{m+k}{r} = \left(\binom{r}{r} + \binom{r+1}{r} + \cdots + \binom{m+k}{r}\right)$$
$$- \left(\binom{r}{r} + \binom{r+1}{r} + \cdots + \binom{m-1}{r}\right).$$

Using the Hockey Stick Identity twice, we see that this is equal to $\boxed{\binom{m+k+1}{r+1} - \binom{m}{r+1}}$.

CHAPTER 14

Exercises for Section 14.4

14.4.1 The expansion is

$$\binom{6}{0}x^{6-0}3^0 + \binom{6}{1}x^{6-1}3^1 + \binom{6}{2}x^{6-2}3^2 + \binom{6}{3}x^{6-3}3^3 + \binom{6}{4}x^{6-4}3^4 + \binom{6}{5}x^{6-5}3^5 + \binom{6}{6}x^{6-6}3^6,$$

which simplifies to $\boxed{x^6 + 18x^5 + 135x^4 + 540x^3 + 1215x^2 + 1458x + 729}$.

14.4.2 $(2y-1)^4$ expands as

$$\binom{4}{0}(2y)^{4-0}(-1)^0 + \binom{4}{1}(2y)^{4-1}(-1)^1 + \binom{4}{2}(2y)^{4-2}(-1)^2 + \binom{4}{3}(2y)^{4-3}(-1)^3 + \binom{4}{4}(2y)^{4-4}(-1)^4,$$

which simplifies to $\boxed{16y^4 - 32y^3 + 24y^2 - 8y + 1}$.

14.4.3

$$
\begin{aligned}
(t^2-1)^3 &= \binom{3}{0}(t^2)^{3-0}(-1)^0 + \binom{3}{1}(t^2)^{3-1}(-1)^1 + \binom{3}{2}(t^2)^{3-2}(-1)^2 + \binom{3}{3}(t^2)^{3-3}(-1)^3 \\
&= \boxed{t^6 - 3t^4 + 3t^2 - 1}.
\end{aligned}
$$

14.4.4 First, factor the expression as $(z^2 + 3z)^5 = z^5(z+3)^5$. Then

$$
\begin{aligned}
(z+3)^5 &= \binom{5}{0}z^{5-0}3^0 + \binom{5}{1}z^{5-1}3^1 + \binom{5}{2}z^{5-2}3^2 + \binom{5}{3}z^{5-3}3^3 + \binom{5}{4}z^{5-4}3^4 + \binom{5}{5}z^{5-5}3^5 \\
&= z^5 + 15z^4 + 90z^3 + 270z^2 + 405z^1 + 243.
\end{aligned}
$$

Thus

$$
\begin{aligned}
(z^2 + 3z)^5 &= z^5(z+3)^5 \\
&= z^5(z^5 + 15z^4 + 90z^3 + 270z^2 + 405z + 243) \\
&= \boxed{z^{10} + 15z^9 + 90z^8 + 270z^7 + 405z^6 + 243z^5}.
\end{aligned}
$$

14.4.5 The term containing x^6 must have 3 copies of $3x^2$ in it, which leaves one copy of $2y$. That term is

$$\binom{4}{1}(2y)^1(3x^2)^3 = 216yx^6,$$

so the coefficient is $\boxed{216}$.

14.4.6 All of the terms are some coefficient times $(x^2)^k(-\frac{2}{x})^{9-k}$, for some value of k with $0 \le k \le 9$. The power of x in this term is $2k - (9-k) = 3k - 9$. We want the term containing x^0, so we solve $3k - 9 = 0$ to get $k = 3$. Therefore, the constant term is the term with $(x^2)^3\left(-\frac{2}{x}\right)^6$. That term is

$$\binom{9}{3}(x^2)^3\left(-\frac{2}{x}\right)^6 = \boxed{5376}.$$

14.4.7 We use the Binomial Theorem for $(x + y)^5$ with $x = 2$ and $y = \sqrt{3}$:

$$(2 + \sqrt{3})^5 = \binom{5}{0}2^0(\sqrt{3})^5 + \binom{5}{1}2^1(\sqrt{3})^4 + \binom{5}{2}2^2(\sqrt{3})^3 + \binom{5}{3}2^3(\sqrt{3})^2 + \binom{5}{4}2^4(\sqrt{3})^1 + \binom{5}{5}2^5(\sqrt{3})^0$$

$$= 9\sqrt{3} + 90 + 120\sqrt{3} + 240 + 80\sqrt{3} + 32$$

$$= \boxed{362 + 209\sqrt{3}}.$$

Exercises for Section 14.5

14.5.1 Apply the Binomial Theorem to $(1 - 1)^n$:

$$(1 - 1)^n = \binom{n}{0}(1)^n(-1)^0 + \binom{n}{1}(1)^{n-1}(-1)^1 + \binom{n}{2}(1)^{n-2}(-1)^2 + \cdots + \binom{n}{n}(1)^0(-1)^n$$

$$= \binom{n}{0} - \binom{n}{1} + \binom{n}{2} - \binom{n}{3} + \cdots + (-1)^n\binom{n}{n}.$$

Therefore

$$\binom{n}{0} - \binom{n}{1} + \binom{n}{2} - \binom{n}{3} + \cdots + (-1)^n\binom{n}{n} = (1 - 1)^n = 0^n = 0.$$

14.5.2 Apply the Binomial Theorem to $(1 + 2)^n$:

$$(1 + 2)^n = \binom{n}{0}(1)^n(2)^0 + \binom{n}{1}(1)^{n-1}(2)^1 + \binom{n}{2}(1)^{n-2}(2)^2 + \cdots + \binom{n}{n}(1)^0(2)^n$$

$$= \binom{n}{0} + 2\binom{n}{1} + 2^2\binom{n}{2} + \cdots + 2^n\binom{n}{n}.$$

Therefore

$$\binom{n}{0} + 2\binom{n}{1} + 2^2\binom{n}{2} + \cdots + 2^n\binom{n}{n} = (1 + 2)^n = \boxed{3^n}.$$

14.5.3   Apply the Binomial Theorem to $(5-1)^n$:

$$(5-1)^n = \binom{n}{0}(5)^n(-1)^0 + \binom{n}{1}(5)^{n-1}(-1)^1 + \binom{n}{2}(5)^{n-2}(-1)^2 + \cdots + \binom{n}{n}(5)^0(-1)^n$$

$$= 5^n\binom{n}{0} - 5^{n-1}\binom{n}{1} + 5^{n-2}\binom{n}{2} - 5^{n-3}\binom{n}{3} + \cdots + (-1)^n\binom{n}{n}.$$

Therefore

$$5^n\binom{n}{0} - 5^{n-1}\binom{n}{1} + 5^{n-2}\binom{n}{2} - 5^{n-3}\binom{n}{3} + \cdots + (-1)^n\binom{n}{n} = (5-1)^n = \boxed{4^n}.$$

14.5.4   Let

$$S = 16\binom{19}{0} + 15\binom{19}{1} + \cdots + (-3)\binom{19}{19}.$$

Note that

$$S + 3\left(\binom{19}{0} + \binom{19}{1} + \cdots + \binom{19}{19}\right) = 19\binom{19}{0} + 18\binom{19}{1} + \cdots + 1\binom{19}{18} + 0\binom{19}{19}.$$

But we know that

$$\binom{19}{0} + \binom{19}{1} + \cdots + \binom{19}{19} = 2^{19},$$

therefore

$$S + 3 \cdot 2^{19} = 19\binom{19}{0} + 18\binom{19}{1} + \cdots + 1\binom{19}{18} + 0\binom{19}{19}.$$

We can now use the identity

$$(19-k)\binom{19}{k} = (19-k)\binom{19}{19-k} = 19\binom{18}{18-k} = 19\binom{18}{k}$$

to rewrite our above expression as

$$S + 3 \cdot 2^{19} = 19\left(\binom{18}{0} + \binom{18}{1} + \cdots + \binom{18}{18}\right) = 19 \cdot 2^{18}.$$

Therefore $S = \boxed{19 \cdot 2^{18} - 3 \cdot 2^{19} = 3{,}407{,}872}$.

Alternatively, we can rewrite S as

$$S = (-3)\binom{19}{0} + (-2)\binom{19}{1} + \cdots + 15\binom{19}{18} + 16\binom{19}{19}.$$

Then

$$2S = S + S = \left(16\binom{19}{0} + 15\binom{19}{1} + \cdots + (-2)\binom{19}{18} + (-3)\binom{19}{19}\right)$$

$$+ \left((-3)\binom{19}{0} + (-2)\binom{19}{1} + \cdots + 15\binom{19}{18} + 16\binom{19}{19}\right)$$

$$= 13\binom{19}{0} + 13\binom{19}{1} + \cdots + 13\binom{19}{18} + 13\binom{19}{19}$$

$$= 13\left(\binom{19}{0} + \binom{19}{1} + \cdots + \binom{19}{18} + \binom{19}{19}\right)$$

$$= 13(2^{19}).$$

So $S = \boxed{13 \cdot 2^{18} = 3{,}407{,}872}$.

Review Problems

14.8

(a)

$$(2x + y)^6 = \binom{6}{0}(2x)^6 y^0 + \binom{6}{1}(2x)^5 y^1 + \binom{6}{2}(2x)^4 y^2 + \binom{6}{3}(2x)^3 y^3$$

$$+ \binom{6}{4}(2x)^2 y^4 + \binom{6}{5}(2x)^1 y^5 + \binom{6}{6}(2x)^0 y^6$$

$$= \boxed{64x^6 + 192x^5 y + 240x^4 y^2 + 160x^3 y^3 + 60x^2 y^4 + 12xy^5 + y^6}.$$

(b)

$$(2x - 1)^5 = \binom{5}{0}(2x)^5(-1)^0 + \binom{5}{1}(2x)^4(-1)^1 + \binom{5}{2}(2x)^3(-1)^2$$

$$+ \binom{5}{3}(2x)^2(-1)^3 + \binom{5}{4}(2x)^1(-1)^4 + \binom{5}{5}(2x)^0(-1)^5$$

$$= \boxed{32x^5 - 80x^4 + 80x^3 - 40x^2 + 10x - 1}.$$

(c)

$$(\sqrt{z} + 2z)^4 = \binom{4}{0}\sqrt{z}^4(2z)^0 + \binom{4}{1}\sqrt{z}^3(2z)^1 + \binom{4}{2}\sqrt{z}^2(2z)^2 + \binom{4}{3}\sqrt{z}^1(2z)^3 + \binom{4}{4}\sqrt{z}^0(2z)^4$$

$$= \boxed{z^2 + 8z^2\sqrt{z} + 24z^3 + 32z^3\sqrt{z} + 16z^4}.$$

(d)

$$(4 - 2x^2)^5 = \binom{5}{0}4^5(-2x^2)^0 + \binom{5}{1}4^4(-2x^2)^1 + \binom{5}{2}4^3(-2x^2)^2$$
$$+ \binom{5}{3}4^2(-2x^2)^3 + \binom{5}{4}4^1(-2x^2)^4 + \binom{5}{5}4^0(-2x^2)^5$$
$$= \boxed{1024 - 2560x^2 + 2560x^4 - 1280x^6 + 320x^8 - 32x^{10}}.$$

(e)

$$\left(x + \frac{3}{x}\right)^6 = \binom{6}{0}x^6\left(\frac{3}{x}\right)^0 + \binom{6}{1}x^5\left(\frac{3}{x}\right)^1 + \binom{6}{2}x^4\left(\frac{3}{x}\right)^2$$
$$+ \binom{6}{3}x^3\left(\frac{3}{x}\right)^3 + \binom{6}{4}x^2\left(\frac{3}{x}\right)^4 + \binom{6}{5}x^1\left(\frac{3}{x}\right)^5 + \binom{6}{6}x^0\left(\frac{3}{x}\right)^6$$
$$= \boxed{x^6 + 18x^4 + 135x^2 + 540 + \frac{1215}{x^2} + \frac{1458}{x^4} + \frac{729}{x^6}}.$$

14.9 Each term of $\left(\frac{x^2}{2} - 3x\right)^7$ is the form $\binom{7}{k}\left(\frac{x^2}{2}\right)^{7-k}(-3x)^k$. The term containing x^{11} is the term with $2(7 - k) + k = 11$, which gives $k = 3$. That term is $\binom{7}{3}\left(\frac{x^2}{2}\right)^4(-3x)^{7-4} = \boxed{-\frac{945}{16}}x^{11}$.

14.10

$$(3 - 2\sqrt{5})^5 = \binom{5}{0}3^5(-2\sqrt{5})^0 + \binom{5}{1}3^4(-2\sqrt{5})^1 + \binom{5}{2}3^3(-2\sqrt{5})^2$$
$$+ \binom{5}{3}3^2(-2\sqrt{5})^3 + \binom{5}{4}3^1(-2\sqrt{5})^4 + \binom{5}{5}3^0(-2\sqrt{5})^5$$
$$= 243 - 810\sqrt{5} + 5400 - 3600\sqrt{5} + 6000 - 800\sqrt{5} = \boxed{11643 - 5210\sqrt{5}}.$$

14.11 Let

$$a = \binom{16}{0} + \binom{16}{2} + \binom{16}{4} + \cdots + \binom{16}{16}$$

and

$$b = \binom{16}{1} + \binom{16}{3} + \binom{16}{5} + \cdots + \binom{16}{15}.$$

We know that $a - b = (1 - 1)^{16} = 0$ and $a + b = (1 + 1)^{16} = 2^{16}$. Thus

$$a = \frac{1}{2}((a - b) + (a + b)) = \frac{1}{2}(2^{16}) = \boxed{2^{15}}.$$

14.12 Consider the expansion of $(x + 2y)^7$. We can find the sum of the coefficients of this expansion by setting $x = y = 1$. So the answer is $(1 + 2)^7 = 3^7 = \boxed{2187}$.

Challenge Problems

14.13 Any coefficient which has a positive power of $(-2\sqrt{z})$ contains a factor of 2. So the only term with an odd coefficient is z^9, and there is only $\boxed{1}$ term with an odd coefficient in $(z - 2\sqrt{z})^9$.

14.14

(a) Using the Binomial Theorem, we can expand $\sqrt{5}$ as follows:

$$\begin{aligned}
\sqrt{5} &= (4+1)^{\frac{1}{2}} \\
&= 4^{\frac{1}{2}} + \left(\frac{1}{2}\right)4^{-\frac{1}{2}} + \frac{1}{2!}\left(\frac{1}{2}\cdot\frac{-1}{2}\right)4^{-\frac{3}{2}} + \frac{1}{3!}\left(\frac{1}{2}\cdot\frac{-1}{2}\cdot\frac{-3}{2}\right)4^{-\frac{5}{2}} + \frac{1}{4!}\left(\frac{1}{2}\cdot\frac{-1}{2}\cdot\frac{-3}{2}\cdot\frac{-5}{2}\right)4^{-\frac{7}{2}} + \cdots \\
&= 2 + \frac{1}{4} - \frac{1}{64} + \frac{1}{512} - \frac{5}{16384} + \cdots.
\end{aligned}$$

Every term beginning with the fourth alternates in sign and decreases in absolute value. Therefore the first three terms determine $\sqrt{5}$ with an error no more than the absolute value of the fourth term. Note that

$$2 + \frac{1}{4} - \frac{1}{64} + \frac{1}{512} - \frac{5}{16384} = 2.2360\ldots$$

and

$$2 + \frac{1}{4} - \frac{1}{64} + \frac{1}{512} + \frac{5}{16384} = 2.2366\ldots.$$

Therefore $\sqrt{5}$ lies between 2.236 and 2.237, so to 2 decimal places, the answer is $\boxed{2.24}$.

(b) Using the Binomial Theorem, the expansion is

$$\begin{aligned}
(x+1)^{1/3} &= x^{1/3} + \binom{1/3}{1}x^{-2/3} + \binom{1/3}{2}x^{-5/3} + \binom{1/3}{3}x^{-8/3} + \cdots \\
&= x^{1/3} + \frac{1}{3}x^{-2/3} + \frac{\frac{1}{3}\frac{-2}{3}}{2!}x^{-5/3} + \frac{\frac{1}{3}\frac{-2}{3}\frac{-5}{3}}{3!}x^{-8/3} + \cdots \\
&= x^{1/3} + \frac{1}{3}x^{-2/3} - \frac{1}{9}x^{-5/3} + \frac{5}{81}x^{-8/3} + \cdots.
\end{aligned}$$

To estimate $\sqrt[3]{9}$, we write it as $(8+1)^{1/3}$, and use the above formula with $x = 8$:

$$2 + \frac{1}{3}\left(\frac{1}{4}\right) - \frac{1}{9}\left(\frac{1}{32}\right) + \frac{5}{81}\left(\frac{1}{256}\right) + \cdots = 2 + \frac{1}{12} - \frac{1}{288} + \frac{5}{20736} + \cdots \approx \boxed{2.080}.$$

14.15 Consider how we compute the expansion of

$$(x+y+z)^n = \underbrace{(x+y+z)(x+y+z)\cdots(x+y+z)}_{n\text{ terms}}.$$

Each term in the expansion of $(x+y+z)^n$ is the product of n variables, and one variable is chosen from each of the n factors. Thus the coefficient of $x^i y^j z^k$ is equal to the number of n-letter strings that contain i x's, j y's, and k z's. So the coefficient of $x^i y^j z^k$ ($i, j, k \geq 0$, $i + j + k = n$) in $(x+y+z)^n$ is $\boxed{\dfrac{n!}{i!\,j!\,k!}}$.

14.16 Let

$$u_0 = \binom{20}{20} + \binom{20}{18}\left(\frac{1}{2}\right)^2 + \binom{20}{16}\left(\frac{1}{2}\right)^4 + \cdots + \binom{20}{0}\left(\frac{1}{2}\right)^{20}$$

and

$$u_1 = \binom{20}{19}\left(\frac{1}{2}\right)^1 + \binom{20}{17}\left(\frac{1}{2}\right)^3 + \binom{20}{15}\left(\frac{1}{2}\right)^5 + \cdots + \binom{20}{1}\left(\frac{1}{2}\right)^{19}.$$

We know $u_0 - u_1 = (1 - 1/2)^{20} = (1/2)^{20}$ and $u_0 + u_1 = (1 + 1/2)^{20} = (3/2)^{20}$. Thus

$$u_0 = \frac{1}{2}((u_0 - u_1) + (u_0 + u_1)) = \boxed{\frac{1}{2}\left(\frac{1 + 3^{20}}{2^{20}}\right) = \frac{1 + 3^{20}}{2^{21}}}.$$

14.17

(a) In $(x + y)^8$, the exponent of y ranges from 0 to 8 inclusive. There are $\boxed{9}$ terms.

(b) In $(x + y)^8 + (x - y)^8$, the exponent of y ranges from 0 to 8 inclusive. All terms are in the form $\binom{n}{k}x^{n-k}y^k + \binom{n}{k}x^{n-k}(-y)^k = (\binom{n}{k} + (-1)^k\binom{n}{k})x^{n-k}y^k$. Since this term is 0 when k is odd, there are only $\boxed{5}$ terms.

(c) In $(x + y)^8 - (x - y)^8$, the exponent of y ranges from 0 to 8 inclusive. All terms are in the form $\binom{n}{k}x^{n-k}y^k - \binom{n}{k}x^{n-k}(-y)^k = (\binom{n}{k} - (-1)^k\binom{n}{k})x^{n-k}y^k$. Since this term is 0 when k is even, there are only $\boxed{4}$ terms.

(d) In any term $a^i b^j c^k$, the exponents of a, b, c sum to 8. So we are looking for the number of solutions in nonnegative integers to the equation $a + b + c = 8$. This is a distribution problem (like distributing 8 pieces of candy to 3 kids), and the number of solutions is given by $\binom{8 + 3 - 1}{3 - 1} = \binom{10}{2} = 45$. Thus, there are $\boxed{45}$ terms.

(e) Just as in part (b), any term in which c has an odd exponent cancels out, so we must only count when c is even. The number of ordered triples of nonnegative integers (i, j, k) that sum to 8 with k even can be counted by cases. If $k = 0$, $i + j = 8$, there are 9 triples. If $k = 2$, $i + j = 6$, there are 7 triples. If $k = 4$, $i + j = 4$, there are 5 triples. If $k = 6$, $i + j = 2$, there are 3 triples. If $k = 8$, $i + j = 0$, there is 1 triple. There are $9 + 7 + 5 + 3 + 1 = \boxed{25}$ terms.

CHAPTER **15**

_____ **More Challenging Problems**

Challenge Problems

15.7 Let the two teams be A and B. We can write the sequence of games as a sequence of letters representing the winner of each game. For example, $BAAAAAAAAAA$ means team B won the first game and team A won the next ten. The sequence ends when the first team has 10 letters in the sequence. However, if we add the rest of the eliminated team's remaining letters to the end of the sequence, we are counting the number of 20 letter sequences with 10 A's and 10 B's. So the number of game sequences is $\binom{20}{10} = \boxed{184{,}756}$.

15.8 A lattice point (x, y, z) is 7 units from the origin if and only if $x^2 + y^2 + z^2 = 49$. We assume $0 \leq x \leq y \leq z$, then

$$3x^2 \leq x^2 + y^2 + z^2 = 49 \quad \Rightarrow \quad x^2 \leq \frac{49}{3} \quad \Rightarrow \quad x \leq 4.$$

Case $x = 0$. We have $y^2 + z^2 = 49$. There is one solution, $(x, y, z) = (0, 0, 7)$.

Case $x = 1$. We have $y^2 + z^2 = 48$. There is no solution.

Case $x = 2$. We have $y^2 + z^2 = 45$. There is one solution, $(x, y, z) = (2, 3, 6)$.

Case $x = 3$. We have $y^2 + z^2 = 40$ and $3 \leq y \leq z$. There is no solution.

Case $x = 4$. We have $y^2 + z^2 = 33$ and $4 \leq y \leq z$. There is no solution.

Now we remove the condition that $0 \leq x \leq y \leq z$ and consider all lattice points. From the first case, we get $3 \cdot 2 = 6$ lattice points, namely $(0, 0, \pm 7)$ and its 3 permutations. From the third case, we get $6 \cdot 2^3 = 48$ lattice points, namely $(\pm 2, \pm 3, \pm 6)$ and its 6 permutations. The final answer is $6 + 48 = \boxed{54}$.

15.9 We will calculate the probability that a random 7-digit number has no ones. There are 9 possibilities for the first digit $(1, 2, 3, \ldots, 9)$, so the chance that it is not 1 is $\frac{8}{9}$. There are 10 possibilities for the second digit $(0, 1, 2, \ldots, 9)$, so the chance that it is not 1 is $\frac{9}{10}$. This is also the case for digits three through seven. Therefore the probability that there are no ones in a 7-digit number is $\frac{8}{9}(\frac{9}{10})^6 \approx .47 < .5$. So the probability that there is at least one 1 $(\approx .53)$ is greater.

15.10

(a) The position of the 1 has 3 choices, while each of the other 2 numbers in the sequence has 2 choices. The answer is $3 \times 2^2 = \boxed{12}$.

(b) The position of the 1 has 4 choices, while each of the other 3 numbers in the sequence has 2 choices. The answer is $4 \times 2^3 = \boxed{32}$.

(c) The positions of the two 1's have $\binom{4}{2}$ choices, while each of the other 2 numbers in the sequence has 2 choices. The answer is $6 \times 2^2 = \boxed{24}$.

(d) The positions of the k 1's have $\binom{n}{k}$ choices, while each of the other $(n-k)$ numbers in the sequence has 2 choices. The answer is $\boxed{\binom{n}{k} 2^{n-k}}$.

15.11 $9 - x$ has the opposite parity (even or odd) as x because the sum of these two numbers is 9, which is an odd number. For every 20-digit number $a_1 a_2 a_3 \ldots a_{19} a_{20}$ such that the sum of the digits is even, the sum of the digits of $a_1 a_2 a_3 \ldots a_{19} (9 - a_{20})$ is odd, and vice-versa. So exactly half of the 20-digit numbers has a sum of digits which is even. Since there are 9×10^{19} 20-digit numbers, half of them, or $\boxed{4.5 \times 10^{19}}$, have an even digit sum.

15.12 Any 2 of the six points determine a chord, so the number of chords is $\binom{6}{2} = 15$. The number of ways to choose 4 chords from 15 is $\binom{15}{4} = 1365$. On the other hand, any 4 of the six points determine a convex quadrilateral, so the number of convex quadrilaterals is $\binom{6}{4} = 15$. Therefore, the probability is $15/1365 = \boxed{1/91}$.

15.13 We will look at the possibilities for the list starting from the end and working backwards. The last element must be greater than all the numbers before or smaller than all the numbers before, so it is either 1 or 8. Similarly, there are two options for the previous element, because it is either the greatest of all of the remaining numbers or the least of all of the remaining numbers. There are two options for all the spots of the list until the first spot, which is simply the remaining number. So the number of ways to order this list is $2^7 = \boxed{128}$.

15.14 The rectangle's first row can be anything from 1 to m, while the last row can be anything from m to j. The rectangle's first column can be anything from 1 to n, while the last column can be anything from n to k. So the answer is $\boxed{mn(j - m + 1)(k - n + 1)}$.

15.15 Let the 9-gon be $ABCDEFGHI$. Fix point A as the first point of the triangle. With equal probability, the second point can either be 1 vertex away (B or I), 2 vertices away (C or H), 3 vertices away (D or G), or 4 vertices away (E or F). If the first two points are one vertex apart, then there is one vertex of the remaining 7 that make up a triangle which contains the center (e.g. ABF or AIE). If the first two points are two vertices away, then there are 2 vertices of the remaining 7 that make up a triangle which contains the center (e.g. ACF or ACG). If the first two points are 3 vertices apart, there are 3 points of the remaining 7 that make up a triangle which contains the center. Finally, if the first two vertices are 4 vertices apart, there are 4 valid points of the remaining 7. Since each of these cases occur with equal probability, the probability that the triangle contains the center is the average of the probabilities of these cases, which is $\dfrac{1}{4}\left(\dfrac{1}{7} + \dfrac{2}{7} + \dfrac{3}{7} + \dfrac{4}{7}\right) = \boxed{\dfrac{5}{14}}$.

15.16 The total number of 3-digit numbers that can be formed using 1 to 5 as digits is $5 \times 4 \times 3 = 60$. Since the number is a multiple of 15, it is a multiple of 3 and 5. Only numbers that end in 5 and 0 are multiples of 5, so the number ends in 5 and is in the form $\overline{ab5}$. Since $\overline{ab5}$ is a multiple of three, $a + b + 5$ is a multiple of 3. The only possibilities for $a + b + 5$ are 9 or 12. If $a + b + 5 = 9$, then (a, b) is either $(1, 3)$ or $(3, 1)$. If $a + b + 5 = 12$, then (a, b) is either $(3, 4)$ or $(4, 3)$. So the four numbers that are multiples of 15 formed with three different digits of 1,2,3,4, or 5 are 135, 315, 345, and 435, which means the probability that this happens is $\dfrac{4}{60} = \boxed{\dfrac{1}{15}}$.

15.17

(a) The cube is divided up into $4 \times 4 \times 4 = 64$ smaller cubes. For a smaller cube to be painted on only one side, it must be on the exterior of the larger cube, but not be along any edge. On each face, there are $2 \times 2 = 4$ smaller cubes that are not on any edge or vertex of the triangle. Since there are 6 faces, $4 \times 6 = \boxed{24}$ of the smaller cubes only have one face painted.

(b) For one of the smaller cubes to have no faces painted, it must be completely in the interior of the larger cube. The interior forms a $2 \times 2 \times 2$ cube within the larger cube, so there are $\boxed{8}$ of the smaller cubes with no face painted.

(c) There are 64 smaller cubes, and $64 \times 6 = 384$ faces of the smaller cubes, each with an equal chance of showing up. On each face of the large cube, there are $4 \cdot 4 = 16$ faces of the smaller cubes painted, and since there are 6 faces, there are $16 \times 6 = 96$ faces painted out of the total 384. So the probability that a painted face comes up is $\dfrac{96}{384} = \boxed{\dfrac{1}{4}}$.

(d) There are n^3 smaller cubes, and thus $6n^3$ faces of the smaller cubes, each with an equal chance of showing up. On each face of the large cube, there are n^2 faces of the smaller cubes painted, and since there are 6 faces, there are $6n^2$ faces painted out of the total $6n^3$. So the probability that a painted face comes up is $\dfrac{6n^2}{6n^3} = \boxed{\dfrac{1}{n}}$.

15.18

(a) Take any 7 numbers from the list of digits $0, 1, \ldots, 9$. If we arrange the 7 numbers in a decreasing order to form a 7-digit number, then we get a number in Rose's list. In fact, we can obtain any number in Rose's list. Thus the number of 7-digit numbers in Rose's list corresponds to the number of ways we can pick 7 digits from the 10 digits 0 through 9. The answer is $\binom{10}{7} = \boxed{120}$.

(b) In a similar fashion, the number of 5-digit numbers formed from the digits 2 through 9 is $\binom{8}{5} = 56$. To each of these 5-digit numbers we append at the end the digits 10. This makes a 7-digit number in Rose's list that has 1 as the tens digit. The probability of selecting one of these is $\dfrac{56}{120} = \boxed{\dfrac{7}{15}}$.

(c) The middle digit can be $3, 4, 5, 6$.

Case 1: the middle digit is 3. The first three digits are chosen from the list $4, \ldots, 9$. The last three digits are chosen from the list $0, \ldots, 2$. There are $\binom{6}{3} \times \binom{3}{3} = 20$ ways.

Case 2: the middle digit is 4. The first three digits are chosen from the list $5, \ldots, 9$. The last three digits are chosen from the list $0, \ldots, 3$. There are $\binom{5}{3} \times \binom{4}{3} = 40$ ways.

Case 3: the middle digit is 5. The first three digits are chosen from the list $6, \ldots, 9$. The last three digits are chosen from the list $0, \ldots, 4$. There are $\binom{4}{3} \times \binom{5}{3} = 40$ ways.

Case 4: the middle digit is 6. The first three digits are chosen from the list $7, \ldots, 9$. The last three digits are chosen from the list $0, \ldots, 5$. There are $\binom{3}{3} \times \binom{6}{3} = 20$ ways.

Note how this confirms our answer to part (a), as $20 + 40 + 40 + 20 = 120$. Now, the sum we are looking for is $(3 \times 20) + (4 \times 40) + (5 \times 40) + (6 \times 20) = \boxed{540}$.

Alternate solution: Given any number in Rose's list, we can pair it with another number in her list by subtracting each digit from 9, then reversing the digits. (For example, the number 9764210 pairs with the number 9875320.) This operation is reversible: doing the same operation to the new number gives the original number back again. So the numbers in Rose's list come in pairs. Also note that the middle digits in each pair always sum to 9. Therefore, the sum of all of the middle digits is 9 times the number of pairs. From part (a), we know there are 120 numbers in the list, so there are 60 pairs, and the sum is $9 \times 60 = \boxed{540}$.

15.19 We will find the probability that the three nickels are drawn in the first 4 draws. Label the coins $N_1, N_2, \ldots Q_5$. There are $\binom{8}{4} = 70$ different combinations of coins that can be drawn. There are $\binom{5}{1} \times \binom{3}{3} = 5$ ways to choose one of the five quarters and all three of the nickels. So the probability of drawing all 3 nickels in the first four draws is $\frac{5}{70} = \frac{1}{14}$. Therefore the probability that the three nickels aren't drawn in the the first four draws is $1 - \frac{1}{14} = \boxed{\frac{13}{14}}$.

15.20 Without loss of generality, let the length of the string AB be 5. Let C, D be points on AB such that $AC = 1$ and $DB = 1$. Let the dividing point be P. The length of longer piece is at least 4 times the length of the shorter piece if and only if P lies on AC or DB. Therefore the answer is

$$\frac{\text{length of } AC + \text{length of } DB}{\text{length of } AB} = \boxed{\frac{2}{5}}.$$

15.21 There are $6^3 = 216$ ways to roll the dice. There are two cases:

Case 1: All 3 dice show different values. There are 6 ways to choose integers between 1 and 6 (inclusive) such the sum of two equals the third and all three numbers are different: $\{1, 2, 3\}, \{1, 3, 4\}, \{1, 4, 5\}, \{1, 5, 6\}, \{2, 3, 5\}$, and $\{2, 4, 6\}$. For each of these, there are $3! = 6$ different ways to order the rolls among the three dice. So there are a total of $6 \times 6 = 36$ ways to roll three different numbers such that the sum of two equals the third.

Case 2: Two of the dice show the same number, and they sum to the third die. There are 3 ways to choose integers between 1 and 6 (inclusive) such that the sum of two equals the third and two of them are the same: $\{1, 1, 2\}, \{2, 2, 4\}$, and $\{3, 3, 6\}$. For each of these different ways, there are 3 ways to order the rolls. So there are a total of $3 \times 3 = 9$ ways.

So the number of ways to roll three dice such that the sum of two equals the third is $36 + 9 = 45$, making the probability that this happens $\dfrac{45}{216} = \boxed{\dfrac{5}{24}}$.

15.22

(a) Annette is the best player in the tournament, so she will beat any player she faces. Therefore she will win all 4 of her games and the entire tournament regardless of the matchups, so the probability that she wins is $\boxed{1}$.

(b) The two participants in the finals are the winners of separate 8-participant single elimination tournaments (the top and bottom halves of the bracket). If Babette is in the same half of the bracket as Annette, Annette will beat her and take Babette's spot in the finals. So the only way that Babette can get to the finals (where she will lose to Annette) is if she is in a different half of the bracket as Annette. When Annette is placed in one half of the bracket, there are 8 spots in the other bracket that Babette can go into, out of the 15 remaining spots, so the probability that Babette is the runner-up is $\boxed{\dfrac{8}{15}}$.

(c) The only people that can beat Colette are Annette and Babette, so for Colette to make it to the finals, Annette and Babette must both be in the other half of the bracket. If we place Annette first, the chance that she is in the other half of the bracket is $\frac{8}{15}$. Now there are 7 spots in the other bracket out of the 14 remaining to place Babette, so the chance that she's in the other half of the bracket is $\frac{7}{14} = \frac{1}{2}$. The probability that they are both in the other half of the bracket and that Colette makes it to the final (where she loses to Annette) is $\frac{8}{15} \times \frac{1}{2} = \boxed{\dfrac{4}{15}}$.

15.23   Let the number of marbles of color red, white, blue, green, be r, w, b, g, respectively. There are $rwbg$ ways to choose one marble of each color; $wb\binom{r}{2}$ ways to choose one white, one blue, and 2 reds; $b\binom{r}{3}$ ways to choose one blue and 3 reds; and $\binom{r}{4}$ ways to choose 4 reds (in all of these calculations we consider choices without regard to order). Therefore, the given conditions tell us that

$$rwbg = wb\binom{r}{2} = b\binom{r}{3} = \binom{r}{4}.$$

From the first equality, we get $2g = r - 1$. From the second equality, we get $3w = r - 2$. From the last equality, we get $4b = r - 3$. So we know that

$$r = 2g + 1 = 3w + 2 = 4b + 3.$$

Therefore, r is 1 less than a multiple of 2, and 1 less than a multiple of 3, and 1 less than a multiple of 4. The smallest such r is 11. If $r = 11$, then $g = 5, w = 3, b = 2$, so the total number of marbles is $11 + 5 + 3 + 2 = \boxed{21}$.

15.24   The terms in the binomial expansion of $(x^2 + \frac{1}{x})^n$ are of the form

$$\binom{n}{k}(x^2)^{n-k}\left(\frac{1}{x}\right)^k = \binom{n}{k}x^{2(n-k)}x^{-k} = \binom{n}{k}x^{2n-3k}.$$

For there to be a constant term, there must be an integer value of k such that $2n - 3k = 0$, which means $k = \frac{2}{3}n$. This can be an integer if and only if n is a multiple of 3. In this case, the constant binomial term is $\left| \begin{pmatrix} n \\ \frac{2}{3}n \end{pmatrix} = \begin{pmatrix} n \\ n/3 \end{pmatrix} \right|$.

15.25 We divide the problem into cases based on when the first game in the winning streak occurs.

Case 1: the first game in the winning streak is the first of the eight games. The last 4 games can be anything. The probability of this happening is $\left(\frac{2}{3}\right)^4$.

Case 2: the first game in the winning streak is the second of the eight games. The first game must be a loss. The last 3 can be anything. The probability of this happening is $\frac{1}{3} \times \left(\frac{2}{3}\right)^4$.

Case 3: the first game in the winning streak is the third of the eight games. The second game must be a loss. The first 1 and last 2 can be anything. The probability of this happening is $\frac{1}{3} \times \left(\frac{2}{3}\right)^4$.

Case 4: the first game in the winning streak is the fourth of the eight games. The third game must be a loss. The first 2 and last 1 can be anything. The probability of this happening is $\frac{1}{3} \times \left(\frac{2}{3}\right)^4$.

Case 5: the first game in the winning streak is the fifth of the eight games. The fourth game must be a loss. The first 3 can be anything. The probability of this happening is $\frac{1}{3} \times \left(\frac{2}{3}\right)^4$.

Thus the chance of a 4-game winning streak occurring is

$$\left(\frac{2}{3}\right)^4 + 4 \times \frac{1}{3}\left(\frac{2}{3}\right)^4 = \frac{7}{3}\left(\frac{2}{3}\right)^4 = \boxed{\frac{112}{243}}.$$

15.26 There are $\binom{12}{5} = 792$ ways to choose any 5 of the 12 CDs. We can make a table of the possibilities where Carlos chooses at least one CD from each category:

Rap	Country	Metal	# of ways	Total
3	1	1	$\binom{4}{3} \times 5 \times 3$	60
1	3	1	$4 \times \binom{5}{3} \times 3$	120
1	1	3	$4 \times 5 \times \binom{3}{3}$	20
2	2	1	$\binom{4}{2} \times \binom{5}{2} \times 3$	180
2	1	2	$\binom{4}{2} \times 5 \times \binom{3}{2}$	90
1	2	2	$4 \times \binom{5}{2} \times \binom{3}{2}$	120

Summing the numbers in the right column, we see that there are 590 ways to select at least one from each category, making the probability $\frac{590}{792} = \boxed{\frac{295}{396}}$.

15.27 Note that of the 27 unit cubes, 8 of them are painted on 3 sides, 12 of them are painted on 2 sides, 6 of them are painted on 1 side, and only 1 (the center square) is completely paint-free. If our initial cube is painted on either 3 or 0 sides, then we cannot possibly roll it so that exactly 1 painted side is showing. So we only need to consider the cases of 1 or 2 sides being painted. If 2 sides are painted, then we win as long as a painted side is on the bottom, which occurs with probability 2/6. If 1 side is painted, then we win as long as an unpainted side is on the bottom, which occurs with probability 5/6. So the probability is

$$\frac{12}{27} \times \frac{2}{6} + \frac{6}{27} \times \frac{5}{6} = \frac{54}{162} = \boxed{\frac{1}{3}}.$$

www.artofproblemsolving.com

The Art of Problem Solving (AoPS) is:

- ## Books

  For over 25 years, *the Art of Problem Solving* books have been used by students as a resource for the American Mathematics Competitions and other national and local math events.

  > *Every school should have this in their math library.*
  > – Paul Zeitz, past coach of the U.S. International Mathematical Olympiad team

  Visit our site to learn about our textbooks, which form a full math curriculum for high-performing students in grades 6-12.

- ## Classes

  The Art of Problem Solving offers online classes on topics such as number theory, counting, geometry, algebra, and more at beginning, intermediate, and Olympiad levels.

  > *All the children were very engaged. It's the best use of technology I have ever seen.*
  > – Mary Fay-Zenk, coach of National Champion California MATHCOUNTS teams

- ## Forum

  As of April 2019, the Art of Problem Solving Forum has over 395,000 members who have posted over 8,100,000 messages on our discussion board. Members can also join any of our free "Math Jams".

  > *I'd just like to thank the coordinators of this site for taking the time to set it up... I think this is a great site, and I bet just about anyone else here would say the same...*
  > – AoPS Community Member

- ## Resources

  We have links to summer programs, book resources, problem sources, national and local competitions, and a LaTeX tutorial.

  > *I'd like to commend you on your wonderful site. It's informative, welcoming, and supportive of the math community. I wish it had been around when I was growing up.*
  > – AoPS Community Member

- ## ...and much more!

Membership is **FREE**! Come join the Art of Problem Solving community today!